John,

Enjoy!

Kee

2008

GREAT RANCHES
OF THE WEST

GR

THE PHOTOGRAPHY OF
JIM KEEN
STORIES BY
JIM KEEN
WITH AMI REEVES

Text and photographs copyright ©2007 Jim Keen

LIBRARY OF CONGRESS CONTROL NUMBER

2007903723

1. Cowboy - Pictorial Works 2. History - Western USA

3. Photography - Ranches

ISBN 978-0-9713355-1-6   Hardcover

ISBN 978-0-9713355-2-3   Limited Edition — Leather Bound

Cover design by Kristopher Orr

Copy editing by

William and Amy Stearns

Proof reading by

Barbara Feuerhaken

Michelle Kahl

Briana Ross

Back cover photograph of Jim on the Salmond Ranch by

William Thomas Gillin

Interior photograph of Jim by

Paul Hensley

Three photographs of the Anderson Ranch

Julia Anderson

Published by

KM Media, Inc

Colorado Springs, Colorado

keenmedia.com

800/363-5336

Printed in Canada

Front Cover Photograph on the Reeves Ranch, South Dakota.

*This book is dedicated to the American Rancher — those hardy people of the land,
who for generations have produced and raised the best livestock in the world.
All of us could well learn from their work ethic, diverse skills and
special character qualities.*

*Also, as in all my work and life, I dedicate this book
to my God — the author of creativity.*

Thanks so Much

There are always so many people who come along side and help on a project like this. First and foremost my appreciation goes to Mike Callicrate, who became aware of the *Great Ranches of the West* project early on. He introduced me to the first ranchers, who own some of the great ranches of the West, and has been a source of information and encouragement through out this project. Barbara Feuerhaken knows these stories as well as I do. Her literary coaching and proofing started with the first written word. Bill and Amy Stearns kept the writing on track. When I wanted to go and tell another interesting story Bill, would guide me back to the main theme. How they found time to think about *Great Ranches of the West* with two newly adopted kids I'll never know. Without Ami Reeves, my stories would have remained good around campfires, but not between bound book covers. Her wordsmithing always improved the stories. When I was at a low point, wondering if I would ever finish this project, the following people were a great encouragement to push on: Jerry and Ingrid Allen, Dennis Allen, Ted and Leslee Bray, Alan and Barbara Feuerhaken, Steve and Brenda Griffith, Bob and Kate Jones, Chet and Sherri Sawyer, Priscilla Sparks, Richard and Sue Michaels, Bob and Jeanette Prather, Dan and Sandy Hagel. My wife, Roann, graciously put up with my traveling from one ranch to another for weeks at a time. She also bought me the first cowboy hat I've owned since I was seven years old. My eagle-eyed proof readers Barbara Feuerhaken, Michelle Kahl and Briana Ross always seemed to find one more typo after I thought it was perfect. Kristopher Orr, my cover designer and layout consultant, went far beyond the call of duty, often meeting me early in the morning before going on to his day job. Thanks, as well, to John Hamilton for designing the "GR" brand that is used through out the book. To the great ranchers who allowed me access to their lives and hosted me during the photography, thank you. My hope is that I have portrayed your lives and history accurately. Thanks to all of you.

TABLE OF CONTENTS

THIRTY GREAT RANCHES OF THE WEST

For additional information about these ranches go online at:
greatrancheswest.com

INTRODUCTION

Towns are dying across rural America today. They are disappearing without even a proper burial. In 1993, there were nearly 900,000 cattle-producing ranches in the country. By 2004, that number had dropped to approximately 775,000. When a ranch fails or goes out of production, it has a ripple effect on its geographic area. The next thing to go is the local hardware store, followed by other small family-run businesses and the local school. Soon, you have a modern day ghost town.

In addition, America's strong, positive, agricultural trade balance is eroding. More and more of our food is coming from other countries and this growing trend has serious national security implications. All of us experience the consequences of our country's dependence on foreign oil, but how much more serious would our situation be if we had a similar dependence on foreign food sources. "We, the people" ought to urge our elected officials and public servants to give due diligence to these matters.

But I am an artist and a visual communicator; I am not a political activist. *Great Ranches of the West* is a celebration of the men and women who produce and raise the best livestock in the world. The thirty ranches represented here are examples of many other great ranches in our country. Throughout this project, I kept facing pressure to add more ranches — there are so many great ranching families and properties. However, I needed to draw the line somewhere or you would need two friends to help you carry this book out of the bookstore. I had to stop with one or two ranches from each western state, a total of thirty.

People have already asked me why such and such ranch is not included here, mentioning their uncle's place or a large corporate ranch. In my definition of a great ranch, I have chosen to focus on ranches that are family-owned and run. I have excluded many fine ranches that are owned by institutions or absentee landlords. Family-run, however, doesn't necessarily mean small, as some of these are the largest ranches in their state. I am not using size as the only gauge of greatness, however. The ranches featured here all have interesting stories and history. The people are men and women of the land, with deep roots. Many have been ranching for five or more generations. They see ranching as more than just a job. It's a 24/7 lifestyle, not something they do for awhile and then move on. One cowboy expressed it like this, "Ranching is a calling. It's similar to a call to the priesthood."

I started working on this project in 2002, and it has been a great education for me, since I have no ranching background. I grew up as a "California beach boy" — living and working most of my life within a short walk or run from the beach. As high school beach bums, all of us locals would shout, "Valley go home!" to anyone we thought was intruding on our beach. Ranchers, forgive me for that youthful, brash attitude towards you. If indeed you then went home and kept your food to yourselves, all of us "cool dudes" would have a hard time surviving on sand and seaweed.

I hope that *Great Ranches of the West* will give all of us an appreciation for these special people and their land. They are tough, well-educated, and independent. They are gifted with a pioneer "can do" attitude. They are ranchers, but also environmentalists, plumbers, EMTs, veterinarians, firefighters, wetland managers, lawyers, equipment repair people, accountants, weathermen, hospitality managers, carpenters, electricians, and heavy equipment operators, sometimes all in the same week. These people are true American heros.

SALMOND RANCH

MONTANA

The full moon rising over the northern Montana Ranch headquarters of the Salmond Ranch.

James and Francis Salmond at their home in Choteau, Montana.

Night after night, sixteen-year-old Libby Smith listened to the cries and screams of fellow prisoners being beaten and killed by their Indian captors. Warriors dragged Libby to the gauntlet on several different occasions. This deadly sport required the prisoners to fight through two lines of men wielding clubs and tomahawks. Few made it through alive. The chief always intervened on Libby's behalf. He'd spared her from death now for five months since her capture and housed her in his own teepee. Would she be the only one not forced to run the gauntlet?

This brave girl learned one thing while witnessing the deadly sport. The strongest warriors stood at the beginning of the line and struck the first, most deadly blows.

Those prisoners who managed to avoid direct hits from the fiercest of the tribe stood a better chance of living through their ordeal. Was this knowledge enough to save her?

Inevitably, Libby's turn came.

That night, Libby hesitated at the front of the gauntlet and turned to the chief with a silent plea. Would he step in as before? This time, Libby sensed how alone she was.

In that split second, Libby noticed that the warriors' attention remained riveted on their chief. Now! And she sprinted. While the gauntlet waited to see whether or not the chief would save Libby, Libby saved herself. The first warriors, caught off-guard, barely had time to ready their weapons before she was gone.

At the end of the gauntlet, a tomahawk blow to the head knocked Libby unconscious. While hovering near death for several weeks, the chief's own daughter nursed Libby back to health, and shortly afterward, a company of soldiers arrived to exchange prisoners with the Indians.

Libby Smith was finally free.

Fifth generation rancher Jeff Salmond, Libby's great-great-great grandson, pauses in his story to point out the lay of the land from our vantage point overlooking the ranch. I'm listening for noises on the wind, waiting with my camera for the full moon to rise. The earth still feels warm under my legs after a dusty, hot July day, but the breeze has turned brisk. I tinker with the tripod, shaking off the chill. Colors of sunset linger on, threaded through the purple of the western evening sky above the ragged Rockies. Over one million-plus acres of the Bob Marshall Wilderness border the western boundary of the ranch—a rich, vast land attracting fishermen, backpackers, and horse riders from all over America. This is wild country.

"So how'd Libby end up here?" I ask.

"I was just getting to that," he says, and tells me that Elizabeth "Libby" Smith was born in Rockford, Illinois in 1845. Ten years after her birth, the family followed throngs of fellow countrymen looking for a brighter future and headed west to Denver. This Colorado boom town, however, proved not to be the land of dreams the Smiths had hoped for. Overflowing with saloons, shabby  businesses, gunfights, and desperate goldminers, Denver provided a rough-and-tumble environment where Libby Smith honed survival skills that would come into play in the future.

At the age of sixteen, Libby and her brother joined

a wagon train heading back east. They planned to visit relatives. It was on this trip that a tribe of raiding Indians ambushed the wagons and took captives, including Jeff Salmond's great-great-great grandmother Libby.

I'm thinking of my daughter when she was sixteen. I take a deep breath, tuning in to Jeff's tale again.

"After her release, Libby reunited with her brother, and the Overland Freight Company employed both of them to accompany wagons from the Missouri River to the Rockies," Jeff explains. The company bosses promoted Libby to a scout, a rare job for a girl. She and her brother built a simple cabin and settled near Virginia City, Montana, where Libby quickly discovered an aptitude for healing when the winter winds of 1863 brought sickness to area miners.

In 1874, at the age of thirty, Libby fell in love with Nathaniel Collins, a silver mine owner and former member of the Montana Vigilantes. Remnants of Nat Collins' colorful past live on at the ranch today where the "77" brand identifies Salmond ownership. "3-7-77" is a figure associated with the vigilantes who cleaned up the lawlessness rampant in 1860's Montana gold camps.

In addition to the main ranch, the Salmonds own this ranch in Eastern Montana where Brent, Lori, Elliot and Emmet Salmond live.

The old ranch homestead at the main ranch headquarters near Choteau.

Nat Collins presented Libby with a cow as a wedding present. The couple settled on 600 acres in this beautiful Teton Valley near the town that would become Choteau, Montana, and began the ranch with 150 head of cattle. Their daughter Carrie was the first white person born in the Valley, birthing a bloodline that courses with determination, survival instincts, and generosity.

Libby became known as "Aunty Collins" in the Valley, selflessly sharing her medical skills with sick or injured settlers and Blackfoot Indians in the area. Libby had come full circle after her capture at sixteen, and became enamored of the Blackfoot, learning their language and customs. She and Nat eventually adopted two orphaned Blackfoot children who, along with Carrie, filled the ranch house with laughter.

When Nat fell ill, Libby took over management of the ranch and the cattle, which now numbered in the thousands. To increase profits in 1889, Libby decided to drive the cattle 90 miles to Great Falls herself and ship them by train to Chicago. When she arrived in town, she found to her shock that the railroad officials refused to let women ride the train with the herd. But the railroad had more fight on their hands than they'd bargained for. For ten days, Libby telegraphed back and forth, battling for the right to accompany her cattle to market like other ranchers. Finally, with the support of her stockmen friends, she won the victory. Railroad officials, fearing a boycott from Montana cattlemen, allowed Libby and her cattle on the train. When she stepped onto the caboose, a cowboy in the crowd waved his hat and whooped, "Three cheers for Aunty Collins, the Cattle Queen of Montana!"

The title stuck, and this compassionate, determined cowgirl became known throughout Chicago and the West as the Cattle Queen of Montana. Libby traveled widely, sharing her ranching expertise and life experience with others.

Jeff Salmond stands on an ancient trail used by the Cree Indians.

In 1904, Libby and Nat's daughter Carrie married Frank Salmond, a prominent rancher in the area, and together the couple continued the family ranching tradition in the bootsteps of Carrie's famous mother. Today Jim Salmond, Carrie and Frank's grandson, oversees the Salmond Ranch operation.

"That's how we got here," says Jim's grandson Jeff, and I see their story spread out before me across the plains and up into the Rockies.

When Lewis and Clark explored the Great Falls Plains southeast of the Salmond Ranch, they expounded on the vast herds of majestic buffalo. Short of the buffalo's decline, this land must look much the same as it did when President Jefferson's Corps of Discovery made their way through the Northwest. I can almost hear the splashes from Meriwether Lewis' panicked escape into the Missouri River with a grizzly at his back. How awed those travelers must have been at the primitive beauty, at the abundance of cutthroat trout, at the endless open skies. This unchanged wilderness continues to tell its stories, both of discovery and of survival, for it takes a certain breed of people to lay hold of this land and cling to it through the generations.

The Cree walked here, too. Winter winds off the Rockies annually urged the Canadian tribe south through Montana to New Mexico. The Cree footpath is still visible across the Salmond Ranch. Some nights, echoes of ghost steps mix with each breath of arctic air. The transition from summer to winter is short and abrupt this close to the Canadian border. Winds off the mountain have been clocked at 100 miles per hour, temperatures can plummet to forty degrees below zero, and snowdrifts high as the house are common.

Jeff breaks into my reverie, telling me how they had to scare two grizzlies off the back porch last May. I ponder the implications of a thousand-pound member of the family Ursidae stumbling over us. This is definitely no place for the timid.

The sky bleeds dark. Now all that's left is for the moon to rise.

Developers and wealthy individuals are snapping up land in this area for second homes, driving the cost of ranching beyond what the average rancher can afford. Many of the oldtimers have been forced to sell out. But the Salmond Ranch, far from average, endures. This family—descended from vigilante/miner, Cattle Queen/Indian captive—clings to this land as if there were no other place on earth.

Because for them, there is no other place on earth, here where the moon is finally rising, full and orange-bright in its summer glory, casting shadows over the Cree footpath worn deep into the Teton Valley. Here where generations of Salmonds have lived and died and marked what providence and determination passed their way with a proud "77."

I snap my shutter, catching this Montana moment.

The Salmond Ranch Remuda (herd) coming back to the corrals

Cattle on the main ranch with the Bob Marshall Wilderness in the background.

Brent Salmond and his two boys look over part of their land in eastern Montana.

From the far upper left: Brent Salmond's oldest boy Elliot, Brent, and younger son Emmet. Elliot is grooming to be a cowboy, and Emmet is a natural mechanic.

Brent Salmond had a serious horse accident shortly before I arrived on the ranch. It broke his arm and punctured his lung, but there was still work to be done. He fought four lightning-caused wildfires on this ranch with his arm in a sling.

Below Left: Lori Salmond leading the horses back from the pasture on the Eastern Montana property.

Below: The tack room on the main ranch is full of saddles won in rodeos by various members of the family.

Far below: One of the hay fields at the main ranch near Choteau, Montana.

In this modern, high-tech world, it seems incongruous that a successful business could still abide by old-fashioned rules like a handshake or a coin toss. But Pete Bonds and family have proven that the simple life can exist harmoniously with the computer-savvy, state-of-the-art, and increasingly global economy.

Two hundred years ago, Comanche Indians hunted plentiful buffalo on this grassland prairie. The tough limestone soil was—and still is—high in calcium and perfect for grazing. Today the Bonds Ranch near Saginaw, Texas, finds itself surrounded by roads. Not your average rocky, pockmarked ranch roads but busy streets and intersections teeming with activity in the form of semi-trucks, construction vehicles, and buzzing commuters. Suburbia is encroaching, an urban world threatening to devour this originally rural area north of Fort Worth. Years ago, World War II American and French-Canadian pilots once practiced bomb-strafing runs over a target on the ranch. It's not much quieter overhead these days, since now there's a busy, private airport across the street, but the ranch continues to fly high.

The Bonds Ranch was established in 1933 by P.R. "Bob" Bonds, who was known for producing top-quality Hereford cattle. When Bob died in 1954, ranch supervisor Pete Burnett began mentoring Bob's youngest son, another Pete. Today Pete Bonds runs the ranch along with his wife, Jo, and their daughters Missy, Bonnie and April.

For Pete, running the ranch is serious big business, so he spends most of his day behind a desk in the headquarters office buying, selling, and negotiating. He's responsible for not only tracking the natural gas operations of hundreds of wells dotting the property, but also for the thousands of head of cattle. These herds are located on eleven different ranches that stretch across a 200-mile radius from the headquarters and are in feedlots as far away as Kansas. Pete says that riding and branding serve as "recreation" for him, getting him out from under piles of paperwork and the demands of his computer. Meanwhile, his three daughters are heavily involved in the cattle operations. In spite of overwhelming workloads, the entire family displays a great sense of humor. Pete's daughters participate

Texas Christian University's Ranch Management Program is one of the best in the country. Once a year they award a set of spurs to a distinguished graduate. Pete received his in the fall of 2006.
The birds gather to watch either the full moon rise or ranch hand Matt Snelus roping calves.

BONDS RANCH

TEXAS

One of Jo Bonds' students getting ready for the Christmas dance recital.

The ladies of the ranch — Jo Bonds, April Bonds, Bonnie Anderson and Missy Bonds.

in everything from roping to vaccinating to castrations to branding, but one of their favorite pastimes is poking fun at the ranch cowboys.

The other ranches Pete monitors often supply extra cowhands for big operations such as inoculating a herd or separating calves from their mothers. One of these cowboys heard a photographer was coming, and he made the ostentatious mistake of getting a sharp new haircut for the pictures before my arrival. Pete's daughters were merciless, calling him "Pretty Boy." Another hand giving Pretty Boy a particularly grievous time roped a calf, and as the calf struggled, his horse stepped over and straddled the rope. The horse then bucked violently, nearly falling back onto the rider. This display of poor cattle-roping then earned the embarrassed man hours and hours of scorn from, of course, the cowboy with the new haircut. And I came away with great pictures of all concerned.

The Bonds Ranch is particularly distinctive for a couple reasons. First, it's the only ranch I visited hosting a full-scale dance studio. Pete's wife Jo teaches ballet and tap to pre-school through high-school students right on the ranch, bringing a dose of culture to this cattle-centric acreage. Second, it's one of only a few ranches where I heard favorable comments about the U.S. Department of Agriculture's NAIS program. The National Animal Identification System tags each cow by fastening a computer chip to the ear which can be read by Radio Frequency Identification (RFID) scanners. The program is designed to track any "mad cow" or other livestock disease outbreak to its source. Early on, the Bonds Ranch adopted the NAIS tagging process, and so far has had no problems or complaints.

So ranching at the Bonds Ranch is happily going high-tech. But does that mean the old-time principles of hard work, perseverance and honesty are on the way out? One night I discussed handshake agreements, people, and rattlesnakes with Pete Bonds.

"I like to deal with honest people," Pete says. "In fact, I'd rather deal with a rattlesnake if I knew it was a rattlesnake than to deal with somebody pretending to be something else but actually is a rattlesnake." Then he tells me a story.

He and a man named Preston quibbled for weeks on the price of a prize-winning, 2,800-pound bull Pete had up for sale. Preston was known for driving a hard bargain, and the two were $16,000 apart in their negotiations. As previous deals with Preston had stretched beyond a two-week span, Pete became exasperated over the current back-and-forth conversations. So he proposed, "Look, Preston, we're not going to haggle over this forever. Let's flip a coin. Heads, we go with my price. Tails, we go with yours."

Missy Bonds' dirty boots reveal who is in charge of the day-to-day ranch work of the cowhands on the Bonds Ranch.

Jo Bonds can "en pointe" on the ballet floor or ride the range with the best of them.

The man replied, "Only if we can use my quarter." Pete agreed, and Preston flipped his quarter.

"Shoot," Preston spat. "You win, Pete. So, okay, I'll pay you the extra sixteen thousand." And so he did.

Pete grins as he adds the punchline to this tale: The coin-flipping conversation took place over the phone. Pete was in Texas while Preston was in Colorado.

Yes, the old ways still guide the principles of the great ranches of the West—regardless of their progress into the 21st century. The Bonds Ranch proves that a culture still exists among these Western operations, a culture that values honesty and the assurance of a man's word, and where a business deal can still be sealed with a handshake.

Pete spends a good share of his day ranching on the computer.

Pete Bonds taking a break from his computer.   *April is busy castrating a young bull.*

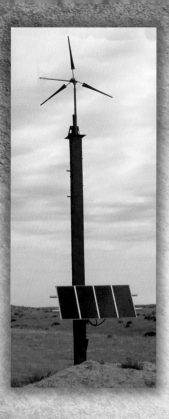

A high-tech solar windmill rises up not far from a fence line and the famous "fork in the road".

James Bledsoe returning from a day's work.

Ranch employee, J. D. Schier and his wife Becky seek a little time alone on the vast high plains.

Fire embers fade, and the cowboy tucks his fiddle and bow away in the chuck wagon. Still, faint music drifts on the wind as one of the two night riders plays his harmonica in the distance. The melody lingers around the 2,000 head of cattle. The other wranglers lie sleeping on bedrolls while the cows rest quietly. Tomorrow morning's early river crossing will be here soon enough, and the cowboys will resume driving the herd across eastern Colorado, bound for Montana.

From the 1860s to the 1880s, cattle camps like this one dotted the Great Plains as up to five million head of cattle were rounded up in Texas and driven to Montana or to various railroad towns along the way. Today, trucks transport the cattle much more efficiently. But the herd still needs to be rounded up, so I'm awake before dawn one chilly fall morning to photograph a roundup on the Bledsoe Ranch on the vast, flat plains of eastern Colorado.

As I step outside I'm surprised by the dense morning haze. It reminds me of the thick, wet fog along the California coast in my boyhood town of Santa Cruz. Sounds are muffled. Mist cloaks the trees around the ranch house until they are shrouded like ghostly sentinels. I hike far out into the pasture and position my cameras on a small berm where the herd will pass by. The fog is nearly impenetrable. I hear men yelling and the lowing of the cattle but see nothing. When you stare into whiteness for an extended period, you develop a sort of visual vertigo. Your eyes play tricks on you; you think you see movement where there is none. It's mystifying how the cowboys can even find the cattle in this fog.

Suddenly there's movement! Cattle are running right for me, less than 25 yards away. Then they turn and disappear. Echoes of the stampede rattle the ground beneath my feet. My heart thumps as I check

BLEDSOE RANCH

COLORADO

my equipment, glad that I'd set up two cameras. Otherwise the action would have faded into the fog before I could say—as my city-bred mind kept humming—"How now, brown cow?"

Just as it does on the California coast, the fog lifts mid-morning, and soon I have as many images as I need of cattle corralled and loaded onto trucks. I put my cameras away and become, as the old hands call it, an honorary CIT—"cowboy in training"—with no idea how much I would learn or how quickly.

I receive my first CIT assignment, that of "gate swinger." To the casual observer, this process of separating calves from their mothers into the numerous pens seems simple. Opening and closing gates, a little noise-making, waving arms and tapping a few stubborn cows on the rear with a bright yellow paddle. After a few gate openings and closings, I have the job down pat. It's time to graduate to something more challenging, something that looks a lot more exciting. I'll be pushing the cattle up the ramp and therefore, I'll need a new title. I promote myself to Assistant Cattle-Ramp Manager. The first few groups of cattle are easily coaxed into the truck. But as the last group mounts the ramp, a steer-sized calf turns around and eyes me. Not to worry, I tell myself. I have a yellow paddle. I wave my arms and swing the paddle, yelling cowboy lingo I've learned. "Back, cow!" But the calf doesn't understand my city accent and charges down the ramp. I resort to the dreaded "tap the cow on the nose" maneuver that I assume stops them in their tracks. The next thing I know, the calf smacks into me like a 450-pound fullback running through the line. Fortunately, the ground below the ramp has been plowed up by thousands of hooves, leaving it soft enough to cushion the blow of my fall. The only relief is that no one captured my embarrassment on film.

While I nurse my wounds and rest for a rare moment, I watch the experts at work. Since 1919, the Bledsoe clan has ranched this region of eastern Colorado, a place where the historic Butterfield Overland Dispatch stage route and the Texas-to-Montana cattle drives once intersected. Oldtimers talk about the Butterfield stagecoach drivers stopping here to allow passengers to watch the massive herds in the distance, and how the travelers would marvel as the wind stroked the miles and

miles of grassland into undulating waves under the high plains sun.

Carl and Josie Bledsoe must have marveled, as well, at the prospects of the land in 1918. With their homeland of West Texas facing a third year of drought, the Bledsoe family loaded their livestock and horses on a train, heading north to a ranch Carl's father had just purchased in east central Colorado. These days, Carl and Josie's grandson, Bill, and his wife, Hilary, run this operation. The ranch, which started out as a basic horses-and-cows affair, is now so much more.

The Bledsoes have developed skills in conservation and management of rangeland and wetlands, and stay up-to-date on financial management and veterinary science. Sons William IV and James are being schooled to continue the legacy of ranching, while daughter Helyna now works with a law firm in San Francisco. Innovators in cattle breeding, the Bledsoe ranch became one of the first in the United States to raise Gelbvieh cattle. Bill also pioneered the conversion of the ranch's power source to solar energy. In the early 1990s he began replacing the old windmills with high-tech solar/wind units. Cattle don't go long without water, and an old broken windmill is only good for photographers who want nostalgic pictures of the Old West.

William Bledsoe roping steers.

With only eight inches of rain a year in eastern Colorado, every drop is treasured.

Of the ranches in this book, the Bledsoe Ranch is nearest to my home in Colorado Springs. So I've returned many times, often taking my grandchildren—who get a kick out of seeing the whimsical metal sculptures on the ranch. The kids compete to see who can first spot "The Fork in the Road," a six-foot-high dinner fork or "Owl Capone," a huge barn owl wielding a Tommy gun to keep pests at bay. In his few spare moments, Bill Bledsoe can be found in his workshop surrounded by chunks of scrap metal, welding art together with humor as a break from the hard work of running a ranch.

My own CIT experiences leave me with a vivid impression of just how difficult it is to run a ranch. That evening I drive west on a dusty road toward home and replay the events of the day, cataloging the little incidents, the sights and smells. My pride and several body parts feel somewhat bruised. As the pickup rounds a corner, I come upon a dozen deer in a draw. A little farther across the plains, a herd of pronghorn antelope race me to the next pasture. In some ways, time stands still in country like this; things are much as they were when the Butterfield Stage lines bounced across the vast plains — remote, unspoiled. I wonder what challenges future generations will face as they struggle to maintain this historic land and lifestyle. I hope the grandchildren of William IV and James will be ranching here like the generations of Bledsoes before them. Yes, most assuredly, I hope so.

Bill in his new tractor.

Lincoln, the pet pronghorn greets visitors to the ranch headquarters.

A foggy morning roundup.

Bill Bledsoe caring for a new calf.

Bill's sister, Sara, rounding up horses.

William Bledsoe, Sr. served for twelve years as a state legislator. He is quick to say that his biggest contribution to the people of Colorado was to keep the government from spending the people's money foolishly.

William, James and helpers branding a calf.

Helyna Bledsoe working on her law school application.

Every successful ranch has someone working in the office. Hilary Bledsoe keeps close check on the income and the outgo.

A well-organized set of corrals makes the sorting of cattle for shipment a much easier job.

Reviewing the day's accomplishments before the long ride back home.

"Boy, you brought this trouble on yourself," the captain shouted, shoving the teenaged stowaway into the arms of two sailors. "Take him below to shovel manure!" It was 1876, and the two rough men hauled the struggling youth to the bowels of the ship. There was no turning back to port for this English cattle ship, so sixteen-year-old Harry Haythornthwaite of Lancaster, England, crossed the Atlantic cleaning the stalls of the bulls bound for Texas. He did his job well, and when the bulls were off-loaded in Galveston, Harry went with them. His care of the imported animals impressed the Texas cattle buyer, who hired Harry on the spot to continue his work with the bulls. Over the next eight years, this Englishman would become a real cowboy, hiring on with cattle drives from Texas to Kansas and the Great Plains. At age twenty-four, Harry pocketed his life savings and tied all his earthly belongings on his horse for one final drive to Ogallala, in the heart of the Sandhills country of Nebraska.

Like other Sandhills ranches, The Haythorn Land & Cattle Company owes its existence today to the early westward fortune-seekers who passed the sand dunes by, unsure how habitable this country would prove to be. With no familiar flora or fauna and the intimidating presence of Native American tribes, only the most intrepid settlers had the fortitude to homestead in the Nebraska Sandhills. Settlers like Harry Haythornthwaite.

Haythornthwaite did what most 19th-century immigrants to America did: He shortened his name to Haythorn, bought a business, and fell in love. Harry and his bride, Emma, operated a livery barn during the early years of their marriage, before deciding to sell the business and hire out to a rancher in the area—she as a cook and Harry as a wagon boss. Harry took his wages in cattle, and in 1905 the couple filed on a section of land east of Arthur, Nebraska, and began ranching on their own. The sand hills that looked so dry and barren to settlers proved to be a key element of the Haythorns' success; the sand hid one of the largest underground sources of water on the continent, the Ogallala Aquifer.

More adventures awaited this hard-working family, as Harry soon threw his saddle onto a train bound for Oregon, where he bought 500 horses and

Top from the left: The Haythorn Ranch is equipped to host weddings, reunions and all types of retreat and business meetings. Craig Haythorn has three generations of silver belt buckles in his family's collection. Craig and Jody relax in what I call a family museum – a room that used to be a barn.
On the right: Christmas lights on the Haythorn Ranch.

HAYTHORN RANCH

NEBRASKA

At the big horse auction on the ranch, bids are accepted from the enthusiastic crowd, as well as bids phoned in and received over the internet.

A good cowboy sometimes needs to get his feet wet to get the job done.

drove them east nearly 2,000 miles back to Nebraska. Meanwhile, Emma handled the ranch. The nearest general store was forty miles away, and Emma would ride side-saddle into town leading a pack horse to buy supplies. She was a crack shot with a rifle, bagging prairie chickens to add meat to the table. These were people of the frontier: tough, resourceful, full of the future.

Five generations later, the Haythorns are hosting a horse auction. My wife Roann and I enjoy a fine afternoon as we stroll among the 2,500 folks milling around the ranch. It's like a county fair—with concession wagons selling food, candy, and ice cream. Old friends congregate around the demonstration arena, where each quarter horse is shown working before bidding begins in the afternoon.

Craig and wife, Jody; sons Sage and Cord, and Craig's mother, Bel, host the auction, where this year 343 quarter horses sell to registered buyers from across the U.S., Canada, and Mexico. The Haythorns run Longhorn, Black Angus, and Hereford cattle on their 75,000 deeded and leased acres, yet the horses are what the Haythorns are famous for. This ranch is one of the largest breeders of registered quarter horses in the U.S. The 500 horses branded with a figure four actually work the cattle, and every cowhand trains his mount for cutting and roping. When the horse is finally sold at auction, the cowboy who trained it shares in up to ten percent of the profits. Craig's late father, Waldo Haythorn, once said, "You can't sell a man's horse and expect him to make another good one overnight."

The Haythorns' involvement in horses isn't just business; they also participate in ranch rodeos across the Midwest and south into Texas. This isn't about the televised professional rodeo circuit, it's about real cowboys riding their own expertly taught horses to compete in events like calf-branding and cutting, wild-cow milking and other events. Since their first year of competition in 1988, the Haythorns have won the all-around high-point award five times at the Abilene Western Heritage Classic Ranch Rodeo in Texas.

It's a working cattle ranch, but it seems to be all about the horses.

As Roann and I wander around at the Haythorn auction, it's obvious that these premier working cattle horses are the stars of the show. But it's the massive Belgian draft horses that fascinate this photographer. The Haythorns decided early on to run the ranch the "old" way—harnessing some of their huge work horses six-abreast to put up 6,000 tons of hay in the summer, pull the chuck wagons during the fall roundups, and haul hay out to the cattle in the windswept Nebraska winter and spring.

When I ask Craig why he uses these horses instead of tractors or other power equipment, he replies, "You don't have to put gas in a horse or change its oil. And it's never too cold to start!"

Above left: Belgian work horses hitched up for a day's work.

Craig Haythorn "cuts" some calves from the herd.

A great ranch runs smoothly with great employees, not all of whom are cowboys. A dependable cook, who provides healthy and delicious meals is a great asset.

My second visit to the Haythorns is in winter, and it's memorable. As I approach the ranch late in the day, the crew of cowboys are driving a herd of running horses across the road and out to a snow-covered pasture. The light is superb, so I pull over and grab my camera, jump a fence, and run to get the perfect picture. Suddenly I plunge through the snow into a ditch running with snowmelt. I'm totally soaked, and it takes two days to dry my boots out — and I didn't even get the photograph.

Yet my frozen feet aren't actually the most memorable part of the visit. The production sale crowds are gone, so I get to spend rich time with the Haythorns and their ranch staff. What strikes me is the span of history, the impact of this family on the world of horses, and also the spirit and culture of this disappearing lifestyle. The Haythorn Ranch is a huge operation, yet it's all about family.

One cowboy says, "Not only are Craig and Jody our bosses, they're our best friends as well." Owners and cowboys alike—some of whom have lived on the ranch for thirty-plus years—work hard together. The schedule is never nine-to-five, and everyone has to be willing to do whatever it takes to get the job done. Even the employees' children get involved in the work. When the job is done, the entire ranch community enjoys socializing together.

Harry and Emma would have been pleased. Like the sand hills on the Haythorn Land & Cattle Company ranch, there's rich treasure hidden just below the surface of this ranching dynasty.

The Haythorn Ranch is known internationally for its registered quarter horses but the ranch is still very much a cattle operation. Several months a year the Belgian horses are used to feed the cattle the old fashioned way and the quarter horses are raised working the cattle.

Joe Bill and Lauren Nunn run cattle on their own land as well as leased property throughout New Mexico.

Below: One of the hired hands brings in a few horses to the ranch headquarters.

NUNN RANCH

NEW MEXICO

I cross the railroad tracks parallel to the two-lane highway, and the road immediately turns to gravel. Soon I pull up to an empty house. It seems that no one is around, so I climb out of my truck, wondering where I should start looking for the owners.

Suddenly a young girl zooms out from behind a storage building on a four-wheel ATV. Sliding to a stop in a cloud of dust, she calls, "You lookin' for Joe Bill?"

"Yes, where can I find him?"

"Follow me," she shouts and takes off before I can get back in my truck. Down the road, I see her waiting for me at a cattle guard. This is Joe Bill and Lauren Nunn's granddaughter Kelsey— the thirteen year old gal who knows how to run everything from the ranch's backhoes to the big semi trucks and also keeps up on horseback with all the cowboys. She is the sixth generation of her family to ranch in New Mexico.

In 1881 David Nunn, his father, and two brothers moved to the little town of Lake Valley in the New Mexico Territory. David tried to homestead nearby, but the Indians drove him off the land. He moved to the relative safety of the town, population about 200. David married Margaret Nunn in 1887, who was no relation. One of his brothers had fallen in love with Margaret's sister and married; and so two Nunn brothers married two Nunn sisters. Like most of the other men of Lake Valley, David went to work in the local silver mine. The success of the mine boomed the town's population to 4,000 residents when silver prices were at their peak in the early

1890s. The Sherman Silver Purchase Act of 1890 required the government to buy millions of dollars of silver bullion annually. But in 1893 President Cleveland repealed the act, and mining in New Mexico slowed to a crawl. Lake Valley shrank to a ghost town.

Over the next several years, David formed a series of ranching partnerships, each larger than the previous one. The last partnership included another senior partner and David's three sons. The ranch stretched over three counties, with 10,000 head of cattle and a thousand horses.

The Nunn sons became involved in the last of the "range wars" of the New Mexico Territory. There was virtually no law enforcement in the Territory in those days, so it was common for a gun to settle disputes. When the "good guys" won a battle, it was another step of bringing peace and security to an otherwise chaotic environment. The boys had constant friction with one particular gang of homesteaders who filled the Nunns' well with rocks and vandalized the ranch property. Hoping to resolve the problems, the three brothers rode out to the homestead. But before they could get close enough to talk, shots rang out from atop the homestead windmill. Two men were shooting to kill. The Nunn brothers jumped from their horses, running for cover. They returned fire, and two men plummeted from their windmill perch. A third gunman hiding out at ground level took a bullet and fell backwards. In an instant, it was over. The Nunn brothers rose from their cover and cautiously approached to check the bodies as an eerie stillness descended over the homestead. All three men were dead. The brothers buried the bodies right where they fell. One of the Nunn brothers was charged with murder, but was quickly acquitted of any wrongdoing.

The next generation saw Smokey Nunn pursuing ranching in the tough new state of New Mexico. When Smokey took over part of the ranch, his wife Eunice would take the train into town for supplies since it was too far by horse, and they couldn't yet afford a vehicle. Their marriage has endured the challenges of ranching for 65 years, and now at age 85, Smokey still gets on his horse to check on his grandkids and great-grandkids as they work on the ranch.

Joe Bill and Lauren keep their ranch separate from Smokey's, but they all work together. Although Joe Bill runs the overall operation, each of the Nunn family members, even the great-grandkids, have their own cattle and brands. The Nunn family now runs cattle on their own land and leased lands throughout the state of New Mexico.

Since the southern boundary of the ranch is only about twenty miles from the Mexican border, the Nunns have had illegal immigrants crossing the ranch for years. The illegal network is now very well developed, with maps that give directions to the ranch water wells. The constant flow of people has actually worn trails across the property.

The family notes that lately there are more and more bad characters smuggling drugs, vandalizing, and stealing along the way. Joe Bill's son Justin recently encountered a lone illegal crossing the ranch down on the southern rangeland. The man was badly dehydrated, so Justin gave him some water and his lunch, telling him to return to Mexico. The next day

Justin Nunn getting the horses ready.

Justin stopped back at his house after picking up the mail—and the illegal traveler was in Justin's house! Fortunately no one was at home, but Justin was unnerved at the thought

Upper left: Even though the sun beats down hot most of the year in southern New Mexico, the winters in the mountains require lots of wood for the stoves.

Far left: Board meetings on the ranch are often held on horseback.

that it could have been his young daughter who had been the first to return to the house. The man escaped into the sagebrush, but the Border Patrol that Justin had summoned soon caught him. Ranching is tough enough without this kind of problem.

And ranching is tough. Joe Bill says, "Ranching is a huge gamble. The profit margins are so small that a slight increase in predator activity can do you in. We have coyotes, bear, and mountain lions. Now the wolves are coming back to attack the livestock. Your cattle can die eating poisonous weeds, or if they don't die you get huge medical bills to nurse them back to health. The biggest uncertainty is drought. A long drought can cause a rancher to cull deep into his herd so that there'll be enough grass for the best cows. It can take years and years to recover if you have to sell a large part of your breeding herd."

Joe Bill's dad, Smokey, the son of hardy pioneers determined to bring order and peace to a wild territory, has seen the ups and downs of the New Mexico ranching world. He has wise words as I'm wrapping up the end of another photography session. "Ranching? It's a series of small crises which you try to catch—before they become big ones!"

Joe Bill Nunn keeps a constant eye on his grass and the weather, moving his cattle often to protect the pasture. Managing the resources cost effectively for a sustained yield year after year is the goal.

Smokey and Eunice Nunn have passed on their wisdom and experience to the upcoming generations of Nunns.

I t's the first day of Spring, and a bitter Wyoming wind whips the falling snow into a frenzy, blowing horizontally into my numb face. The horses stamp their feet in anticipation; their breath fogs the dark morning air. Twelve cowboys and one cowgirl work proficiently in the cold, cinching saddles tight, talking excitedly about the 20-mile cattle roundup.

My hands are so stung with cold that I'm having difficulty changing lenses on my camera. I mention to one of the guys that the wind feels pretty icy for a Spring day.

"Jim," he says, grabbing my shoulder and looking me straight in the eye, "it doesn't get any better than this!"

I turn my attention back to my uncooperative camera and wonder how many people in other lines of work would get up early in this kind of weather and love it. But Wyoming's Warren Ranch has always attracted owners and hands with an abundance of enthusiasm, work ethic, and respect for the job. For a year and a half now, I've been chronicling Warren Ranch through photographs that portray this lifestyle in all seasons of the year. I've acquired a deep appreciation for the hard work of these ranchers and the close association that they share with the land.

As the group rides away from me, the rich history of the ranch echoes under the staccato of horse hooves like far-off gunfire. The tale of the Warren Ranch begins miles and years away in Port Hudson, Louisiana, as a nineteen-year-old corporal, eager to speak to his commander, threaded his way through the medics tending fallen men. The rumble of cannon fire faded as darkness fell. Francis Emory Warren of Company C intended to volunteer for a dangerous mission to lead an early advance against the enemy in the morning. For personal bravery during that maneuver in May of 1863, Warren earned the Congressional Medal of Honor.

The Civil War did much to shape the character of this young man from Hinsdale, Massachusetts. After the war, the lure of the Western Territories called Warren away from Hinsdale to the wild, boisterous town of Cheyenne, Wyoming. Warren found work with cattle baron Amasa Converse, and their arrangement evolved into a partnership that ultimately formed the Warren Livestock Company.

Warren's business and ranching pursuits prospered from the beginning, quickly

An eagles nest at dawn.

WARREN RANCH

WYOMING

followed by an interest in politics. He was elected first governor of Wyoming and then served as a United States Senator for 37 years.

Innovative in his ranching techniques, Senator Warren incorporated practices still in use on the ranch today. Almost from the beginning, Warren Livestock Company ran both sheep and cattle. Cattlemen and sheepmen clashed in those early days, an animosity that often reached violent levels on the range and dictated a separation of the species. But here on the Warren Ranch, sheep and cattle graze the same pastures, creating an environmental advantage when the sheep eat down noxious weeds, leaving the more nutritious grasses behind for the cows.

Senator Warren's son Fred brought Harvard-trained engineering experience to the ranch, moving it from its "Wild West" phase into an efficient ranching affair. Fred updated facilities and equipment until it became one of the most modern operations in the West. Fred also worked with Dr. John Hill of the Wyoming School of Agriculture to develop the Warhill Sheep, a breed with a natural tendency to twin and well suited to a range environment.

When Fred died in 1949, son Francis E. Warren III assumed ranch management duties, but fourteen years later he sought an outside buyer because he had no heirs. Coloradan Paul Etchepare acquired full ownership in 1973, with plans to develop some of the range acreage as cultivated cropland. By 1992, the Etchepare family had 12,000 acres of farmland. The ranchers used the remaining stubble from every harvest as feed, increasing an overall livestock-carrying capacity.

Closing the gate on the first day of Spring.

Current owner Doug Samuelson continues the fertile traditions of this ranch. He works tight alongside the hired hands except at the beginning of the year when he serves in the Wyoming State Legislature. Doug applies excellent wetland and grassland management practices to the immense wide-open spaces under his care. A great diversity of wildlife coexists with the livestock at the Warren Ranch. Some migrate with the seasons, while others remain as constant companions of the sheep and cattle.

I've been meaning to travel to the summer pasture—about 50 miles from ranch headquarters—and photograph the sheep, so later that year, my wife Roann and I pick up the gate key for the most remote part of the ranch. I leave a voicemail with the sheep manager for directions to the pasture, knowing full well that I'll never be able to get a return call out there where there are no cell signals. Roann and I head out anyway.

After the snow melts in the mountains near Laramie, the Warren Ranch trucks the sheep out to the summer pasture in the high country. It's steep, rocky terrain accessible by foot or, in our case, a trusty four-by-four.

Above: The cattle are following the fence line as they are driven to fresh pasture.
Other than the companionship of their sheepdogs, sheepherders are alone most of the time — just the sheep, the dogs and the Wyoming wind.

After about thirty minutes we come to an old unused corral and a couple decomposing cabins. As soon as I can load the cameras, I'm out of the truck photographing the weathered wood. Afternoon light glows across the old boards, giving them a deep, luminescent texture. I'm deep into the creative process, so I hardly hear the bleating of far-off sheep until Roann brings it to my attention. High above us, a few animals meander along the ridge. We pack our equipment and start climbing.

Long before we near the herd, a Great Pyrenees dog leaps toward us down the hillside, barking with alarm. Each sheepherder on the Warren Ranch has a

couple of Border Collies to help move the 800-1,000 sheep, as well as a large guard dog for protection. For the next hour or so, we continue scrambling up the incline and the Great Pyrenees travels alongside, always remaining between the sheep and us. We never find the sheepherder, but most likely he doesn't care to be found. Most of the reclusive sheepherders on this ranch are young men from South America, here on ten-month work visas. They only see another human when the sheep manager stops by for weekly visits to restock provisions. On the way back down to our truck, we come across a clearing with a fire ring, obviously the sheepherder's campsite from the previous night. In the fading light, Roann fixes dinner on the camp stove while I unload film and clean the cameras. We crawl into our sleeping bags in the truckbed and count meteors and satellites, pointing out the occasional shooting star before the big sky absorbs the star's trail. The faint bleating of sheep carries us into sleep.

Far from the lights of cities the moon seems so much closer and brighter.

Doug Samuelson keeps a close eye on his vast cattle and sheep operation. The ranch has its own breed of sheep – the Warhill variety. Ever since its founding by Governor Warren, Wyoming's first governor, the ranch has run both species of livestock.

A new-born lamb.

Doug Samuelson, his ranch manager, Steve, and Steve's son check on a water source.

A stream flows past the JD Sheds, a remote lambing area north of the ranch headquarters.

The century-old tale has been passed down through six generations of ranchers, undiluted by time. In 1893—fifteen years before Oklahoma would become a state, Oscar Chain traded $50 and a shotgun for 160 dusty acres situated between the North and South Canadian Rivers in Dewey County. Looking over the ranch today, it seems to me that the founder of the Chain Land and Cattle Company got a fabulous bargain. With 60,000 acres spread over seven different properties from Oklahoma to Kansas, the ranch has diversified into so many different markets that Oscar Chain might not even recognize it any longer. But things haven't always been so rosy.

"One day the ranch accountant burst into my office," says Ralph Chain, chuckling over the story. "He told me things were so grim right now that my granddad got a bad deal." At eighty-something years of age, Ralph can remember most of the hardest times, times when the land didn't seem worth even fifty bucks and a gun. He recalls the dust storms that chewed up the prairies and spit them back out, unrecognizable. Ralph's wife Darla can remember her mother's stories of those dusty years. Darla was just an infant when her mother would moisten baby clothes and hang them over Darla's face to keep her lungs clear and keep her alive.

Ralph and Darla have never forgotten the dustbowl lessons of how fragile this land is. Before Ralph took over ranch operations from his father, wildlife had been scarce for the previous sixty years, killed off by homesteaders and unstable environmental practices.

Above: Grandson Newley Hutchison in a hurry on his horse. Ralph Chain in the hunting supply store on the ranch. Crossing a stream on one of the Kansas ranches. The ranch headquarters in Oklahoma.

CHAIN RANCH

OKLAHOMA

Now whitetail deer, birds, wild pigs, and other creatures native to this Oklahoma region flourish where the Chains have taken the time to create specific habitats for them on the ranch. "My goal is to put this country back together the way God made it," Ralph says. His cell phone interrupts our conversation. I scribble some notes and smile to myself at the ring tone that announces a call on this rancher's phone—a rooster crowing!

A crowing rooster is an apropos jingle for a man who is up and around at the crack of dawn to oversee a conglomeration of ranches stretching across two states. He doesn't golf or travel in his free time; his hobby is his ranch, especially when he's creating new ponds to entice wildlife or coddling pastures to make them more productive. Ralph spends hours on his bulldozer, working with the land instead of against it.

"I want to make the land better now than when I got it from my father," he says after finishing the phone call. He indicates the surrounding meadow and tells me that hundreds of acres of cultivated land have been returned to natural grasslands where the cattle graze on a rotating basis. From where we stand, I can see the fence line dividing their property from a neighboring leased property. Even my untrained eye notices the huge contrast between a well-managed pasture and one that has been neglected.

Darla and Ralph have carefully restored period details on the ranch, including a one-room schoolhouse complete with old desks—as if the interior

The Chain Ranch is a diverse family operation spanning six generations.

is simply holding its breath until it hears the rustling of students sliding into their seats. There's an authentic log cabin and an historic Sears & Roebuck home—ordered from the famous catalog and assembled on-site in 1910 from pre-numbered parts.

Cattle is big business at the Chain Land and Cattle Company, with 2,500 females bred each year and calves raised on natural grass. The Chains' beef is marketed

Roger Van Ranken manages one of the other Kansas properties.

The Chains have some great employees that have worked for them for twenty years or more. Mike Jones manages one of the ranches in Kansas, where the registered quarter horses roam.

Mike restored the old chuck wagon you see on the right. They use it on the Spring roundups.

through the all-natural Coleman Natural Beef brand out of Denver, a supplier of hormone-free and antibiotic-free beef.

The Chain Ranch has not only grown from 160 to over 60,000 acres, but it's also diversified. I've seen a unique breed of longhorn cattle roaming the ranch. There are also quarter horses, a hunting lodge, a hunting club, fishing opportunities, and guided hunts for a wide variety of wildlife. A native grass seed company comes in every year to harvest between 2,000 to 3,000 acres of the grass that has been allowed to thrive. The ranch is also a regular stop on a tour for elementary teachers to see how a real ranch operates, so they in turn can educate their students. Because of the Chains' outstanding land management, they were selected by the National Cattlemen's Beef Association for a 2004 Regional Environmental Stewardship Award.

But the Chains are not only good stewards of the land. They're also known as honest and upright business owners, the likes of which are not often found these days. Ralph tells me a story that illustrates this point perfectly.

For seventeen years, Ralph leased a ranch from a prominent Chicago attorney who didn't know the first thing about Oklahoma land pricing. He was so green "he couldn't tell a trail from a tumbleweed," Ralph says. Ralph called the attorney one day and informed him that the property lease was too low in comparison to the going rates. The lawyer was astounded to find a lessee who would offer a higher payment. This act of forthrightness stayed with the Chicago lawyer. Many years later, Ralph let the attorney know they'd no longer need to lease the land since they were purchasing acreage on the other side of the leased property.

The lawyer, who by this time was advanced in age and had no heirs, began to cry. "You can't do that," he told Ralph. "I'm going to price this land so low that you can't afford not to buy it over the other property!" He was so impressed by the work ethic and honesty of Ralph and his family that he sold the ranch to the Chain Land and Cattle Company for a song.

From the 1880 cattle drives that wound through this Cheyenne Indian Territory to the devastation of the Dust Bowl days, the Chain Ranch has emerged from history stronger, bigger, and better. The thing that will stay with me most from this trip is the pure beauty that such hard work has wrought. The Chains have stayed true to the land, and it has paid them back a thousand-fold.

You couldn't pay me $50 and a shotgun to turn away from these sights and sounds: the nodding alfalfa as the wind rakes through the grasses, the longhorns grazing in the next pasture, and the cooing of doves as they call from the horizon. My camera records the images; my mind records the smells and sounds. The legacy of the Chain Land and Cattle Company can be found in the land itself.

There are two types of cowgirls on the Chain Ranch – one wrangles paper rather than horses (Bobbie Parall wears the hat and Linda Fields carries the ledger books.).

Beginning in the 1960s, Ralph Chain has created ponds and natural corridors on the ranch to attract more wildlife. In 2004 the ranch was named a regional winner of the Environmental Stewardship Award of the National Cattleman's Beef Association. As part of wise pasture management, about once every five years, when weather conditions are right, fire is used to restore balance to the rangeland.

1931. The air stood still. But across the cobalt Kansas sky a 200-foot-high wall of black clouds advanced like a biblical plague, blowing dirt with gale-force winds the Easterners politely called "dust storms."

Charles Duff was nine years old when his parents packed up their 1925 Model-T with four kids, a cat, a dog, and a duck and headed for the Kansas plains. Charles remembers riding in the back of the truck with no worries about summer showers because it hadn't rained for two years. He says they arrived lighter than they started out, with four sunburned kids, a dog, a duck, two eggs, and no cat.

People all across the country were on the move. During the Great Depression's second year and a summer of record high temperatures, the effect of three thousand bank closures had rippled out from Wall Street to the Great Plains. One out of three farmers on the plains faced foreclosure. Then came the dust storms that Charles recalls so vividly.

The spry eighty-year-old tells me about the first storm as I scribble notes. He had just run out of the schoolhouse one day when a sky choked with mountains of black clouds grabbed everyone's attention. When the wind finally hit, dirt blocked out the sun until it was, as Charles says, as dark as the inside of a horse. He couldn't see his hand in front of his face. The dust had a coarse, sharp edge that chafed any exposed skin. He clung to a fence to keep from being blown across the schoolyard.

The storms came more and more often. His mother would race to soak blankets and sheets with water, hanging them over the windows and doors to keep the grit at bay. Then, after the maelstrom, the family would literally shovel the dirt out of the house. To this day at Charles' medical checkups, the doctors say, "You went through the Dust Bowl years out on the prairie, didn't you?" once they see the damage to his lungs from so much dirt.

But it wasn't just billowing clouds of soil. The wind brought other plagues. Grasshoppers swooped down like an army from the north until they blackened the ground, stripping every green leaf on the few, sparse trees and devouring all the grass they could find. Insects crunched underfoot. The ruthless grasshoppers squeezed into houses, hopping across kitchen counters and into beds. When they finally moved on, the average Kansas homestead looked as if it had lost a war, skeletonized by an insatiable army. Ninety percent of the crops that managed to survive the drought ended up being eaten by the grasshoppers.

Duff Buffalo Ranch

KANSAS

Even more unsettling were the years of the rabbits. Hundreds of thousands of rabbits crawled across the plain, consuming everything in their path. Charles vividly remembers a school bus pulling up outside the school to recruit boys for rabbit hunting. Each boy received a baseball bat or a club. Grown men would form a long line in the shape of a "V," with the boys in a second row behind them. Though it seems unsavory in today's environment, the men began clubbing rabbits to death for the sake of their families' survival. A typical hunt would kill three to four thousand rabbits in a few hours. The boys would then return to school while the men would stuff the carcasses into 50-gallon drums and load them on trains going east to feed mills. These rabbit kills happened over and over again in towns all across Kansas, Oklahoma, and parts of Texas, Colorado, and New Mexico.

According to Charles, the first eight years of the 1930s were "the worst hard time," just like the title of Timothy Egan's recent Dust Bowl book. I have no trouble believing him.

It was a stark era for what is now the Duff Ranch near Scott City, Kansas. A little more than a century before this, however, these plains were covered with rich, hardy prairie grasses and grazed by massive herds of animals as tall as six feet high, weighing nearly a ton, and able to sprint at thirty miles per hour from a dead stop: the American buffalo. There were seventy-five million buffalo in America when, in 1806, explorers Lewis and Clark wrote: "The moving multitude darkened the whole plains." Most tribes

of nomadic Native Peoples followed the herds' migrations and used every part of the animals for food, clothing, materials for shelter, and even sinew bowstrings.

By 1800, the small buffalo herds east of the Mississippi River had disappeared. Yet millions roamed the Great Plains until buffalo hunting became the prime sport of this 1830s frontier.

It was in 1868 that a buffalo hunter named William Cody contracted to provide meat for the twelve hundred men building the Kansas-Pacific Railroad. A former Pony Express rider and scout for General George Custer, Cody was challenged for the title of "Buffalo Bill" by another Kansas buffalo hunter from Fort Wallace named William Comstock.

Near what is now the Duff Ranch, Cody and Comstock competed in a day-long "Champion Buffalo Hunter of the Plains" contest. Cody won 69-48, and thereafter was known worldwide as Buffalo Bill. In 1882 he launched his "Buffalo Bill's Wild West Show" featuring legends such as Annie Oakley, Sitting Bull, Red Cloud, and Frank North. Touting re-enactments of Custer's Last Stand, the show toured the East Coast, England, France, Spain, Italy, Germany, Austria, and Belgium. The size, weight, and aggressive disposition of the buffalo ensured that none of these beasts actually appeared in Bill's Wild West circus.

Herding buffalo is dangerous business – they are big, strong and fast. Sometimes several will find their way down into the canyons on the Duff Ranch. Getting them out requires all the knowledge and resourcefulness Richard Duff can muster.

One of the original rock buildings is still standing on the ranch. While the buffalo roam, Charles Duff keeps accurate accounts in his office.

This is a rare, off-color pair of buffalo with some albino characteristics.

A by-product of ethanol production from corn is used as feed for both buffalo and cattle at the Duff's feed lot.

When the rains come, it kisses the grass. Some of this native grass is harvested for the seeds and sold for a premium price.

In fact, by 1900 no buffalo appeared on the Great Plains either. The indiscriminate slaughter of buffalo for recreation left fewer than 300 in all of North America. Fortunately, federal legislation began protecting these unusual animals. Today, after a century of conservation and private breeding, the total number is back up to 200,000. Six hundred of them reside on the Duff Ranch.

Charles Duff's youngest son, Richard, became fascinated with buffalo during his Native American studies at college about 40 years ago. Visiting a Sioux community in South Dakota, he returned to the Kansas ranch with a pair of the beasts. "I would just sit and watch them out the window," he says. The Duffs' buffalo run on 9,000 acres, while the ranch also keeps up to 8,000 cattle in their feed lots.

Buffalo require a special alertness and special constraints. They're fast sprinters, and can surprise you in amazing ninety-degree turns by jumping straight up and switching direction. Fences must be built at least six feet high. The beasts are too ornery to be herded by horses, so the Duff Ranch crew uses pickups. My advice after observing some buffalo-herding: Never buy a used pickup from a buffalo ranch!

Buffalo are not domesticated. They're still wild. Especially in mid-summer mating season, the herds become restless and quarrelsome, with the huge bulls pawing the ground in combat. One summer, Richard Duff carefully separated one of these massive animals from the others, using a BB gun to urge the beast into the pen. He carefully closed the ten-foot gate with its three-inch-thick, four-inch-high hinges, and then looked back over his shoulder to see how the other penned buffalo were reacting. The buffalo he'd just corralled leaped high in the air, smashing into the gate and ripping it off its hinges. As the huge gate fell, Richard was crushed underneath. The cowhands thought he was dead, but it's hard to keep a Kansas man down. After being unconscious for 20 minutes with a severe concussion, Richard was very much alive—though his back now gives him some trouble.

Why do the Duffs bother with hundreds of cantankerous buffalo? Because, just as the Sioux and other Native Peoples found, they're worth the challenge. The heads and hides still command excellent prices. The meat is lean—higher in protein than that of any domestic animal—and natural with no added hormones or antibiotics.

Some tribes revered the buffalo for the good fortune they brought. It seems to me that good fortune has found a permanent home at the Duff Ranch, whose founders fought back at those blizzards of displaced topsoil until their very bodies were marked by the weather. Refusing to be kept down, refusing to abandon their land, the Duff family, much like the buffalo, has returned from the brink of extinction to become a thriving, flourishing part of the Kansas landscape.

Don Flournoy

Likely, California, population 200. The town may be small, but it's even smaller from my airplane window, where miniature buildings and ant-sized vehicles continue to shrink with each dizzying foot we climb. I try to relax a white-knuckled grip on the arm of the passenger seat of John Flournoy's vintage 1952 light plane.

I laughed earlier when John asked if I was sure I wanted to go up to take aerial photographs of the ranch. "The last plane we had crashed, you know. But then, it was much older. Still, some folks—." The roar of the antique engine had drowned out John's voice for a moment. "Some folks blamed it all on pilot error," he yelled. Now we swing over rich, green Modoc County, tilting so I can get a good shot. I'm trying to figure out if he was kidding about that crash story. Actually, John is a thoroughly expert pilot whose civilian skills were once put to good use training fighters for combat in Vietnam.

The plane, old as it is, is impeccably maintained by the mechanic in the family, Dave Flournoy. Yesterday I asked Dave about the huge equipment graveyard on the ranch, a section of old, broken tractors and implements that's a familiar sight on almost all ranches.

"Graveyard!" he retorted. "That's no graveyard. That's our parts department!" Dave maintains all the ranch equipment, the plane, and even his dad's classic 1954 pickup purchased decades ago for $500.

The resourcefulness and plain hard work of the Flournoy family is balanced with a kind of California cool. It's obvious when I ask Dave about the ranch's Lazy F brand, a good, strong "F" relaxing on its side. "Well," Dave says, "There's always the choice between making a good living or making a good life."

It was a good life that John D. Flournoy was searching for when he first rode into this high country in 1871. One of the thousands of young Missouri men drawn west on the Oregon Trail, Flournoy wandered into California and worked as a cowhand in Yolo County in 1865. When feed provisions fell drastically low in Yolo County as the winter of '71 approached, John saw an opportunity. He and a partner offered to winter ranchers' cattle up in the lush grass of the 4,000-foot elevations of Modoc County. The venture yielded resounding success when they drove the fattened, healthy cattle back down to Yolo County in the spring. So the next year, the partners bought all the cattle they could—600 head—and herded them up to the grasslands near what is now the town of Likely. Unfortunately, Flournoy didn't realize that the previous mild winter had been a fluke of nature. The winter of '72 came early and in full force. In the hard, cold weather, all of the cattle died except one. Flournoy and his partner kept the lone steer

Bill Flournoy — the cowboy.

John Flournoy — the farmer.

Dave Flournoy — the mechanic.

LIKELY LAND & LIVESTOCK INC.

CALIFORNIA

alive through the bitter winter by feeding it bark they stripped from willow trees.

But John Flournoy survived and still remained enchanted by the high country. He married in 1878, taking his bride Mary on a 450-mile honeymoon buggy ride around Modoc County to thoroughly explore their new property.

The Flournoys were good at raising cattle, driving them down to Yolo County for the stock sales. John would then brave the dangerous trails back up to their ranch—dangerous because local bandits knew that ranchers were paid in gold dust.

The area had always been hazardous. General John "Frontier" Fremont and his trusty guide, Kit Carson, were the first white men to explore this country in the 1840s. The native Pit Indians defended this region against would-be settlers until about the time John Flournoy put down roots here. Several hundred Pit Indians still live in the area. Don Flournoy, the 89-year-old patriarch of the Flournoy family, had childhood Indian friends who taught him their language, and he has not forgotten. Don is probably the last remaining speaker of the Pit dialect.

Don is retired but still has a home on the ranch. He shows up every morning for the 6 a.m. breakfast meetings, along with six or seven full-time hands.

I'm impressed by everyone's punctuality—until I dig into one of the finest breakfasts I've had on any ranch in the great American West. You'd have to be crazy to risk being late and missing this feast!

While I'm here photographing the Likely Land and Cattle Company, I'm staying in the bunkhouse, a separate building next to the cook's galley. She bakes fresh, homemade bread at 4:30 each morning, and then whips up a mouth-watering breakfast, all of it hitting the tables by six. The big noon meal is just as delicious, drawing everyone in from all over the ranch. Some of the hands have worked here nearly thirty years, and I'm thinking the food has to be one of the compelling reasons. Friends and associates outside the ranch know that noon is the time to catch all the busy Flournoys at once, so phone conversations punctuate the entire

When it is time to round up cattle, neighbors and the wives of employees all saddle up.

It's early Fall on the Flournoy Ranch.

meal.

Don has turned over all aspects of the ranch business to his three sons. Dave, the mechanic, is the youngest. He tells me he's "in charge of sixty-four radiators." My new pilot friend, John, handles the farming and irrigation. The ranch puts up between 4,000-6,000 tons of hay annually, enough to feed all their own livestock and still have some left to sell each winter. The ranch sets aside 3,000 acres for farming, with fields laser-leveled for irrigation efficiency.

To me it's a photogenic junk yard of old cars and tractors, but they call it their parts department.

Bill, the oldest, is president of the family's corporation and oversees cattle operations. He's so focused on cattle that he figures he spends three hundred days a year in the saddle. Riding isn't quite as pleasant as it used to be for Bill, since he was recently bucked off a horse and broke his pelvis. The whole family is proud of Bill's business acumen. They point back to the year that he turned twenty, when his father Don departed for an extended five-month trip and left the ranch in Bill's hands. The young man had to learn how to run everything.

Don also talks about a time he and young Bill went into town to the bank to make a loan payment. "How much do these tellers make, dad?" Bill asked. Don told him. Later Bill asked, "How about the bank manager?" Don guessed. On the way back to the car, Bill said, "I just figured out that the interest we're paying on that loan is enough to pay the salaries of everybody who works in that bank. Why don't we sell some cows and pay off that loan?" And that's what they did.

Together the Flournoys look to the future. They're relentless in their well-organized, and hard-working farming and the cattle business, but they predict that some of the ranch's future may be in its water. The Lazy F has springs and wells scattered across the property, and the corporation owns substantial water rights to the South Fork of the Pit River. At 4,000 feet in elevation, the growing season is only 100 days long, while down in Yolo County and the greater Central Valley of California, it's double that. That extended growing period seems ideal except for the lack of sufficient water during these months. "We might find that selling water is even more profitable.

Helen Popp, the Australian wrangler, is picking out her horse for the day.

The roundup begins with a ride through the sage brush, progresses to a gravel ranch road, crosses a highway and catches a stray motor home.

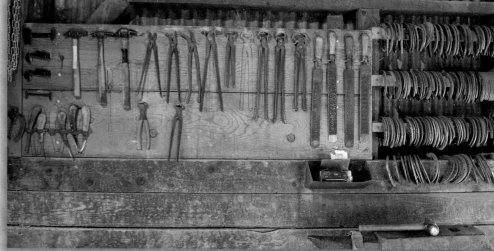

Bill Flournoy's wife, Athena, has a house full of interesting collections.

Bill's barn has its share of collections as well,
all of the working variety.

Athena has an office with a view.

This is what mechanics do when there is nothing broken.

Phillip Hamnest

During the migration season, these wetlands on the ranch attract thousands of birds.

Lance Froelich leads the quarter horses across the Cannonba

FROELICH RANCH

NORTH DAKOTA

✳

River on their ranch near the Standing Rock Indian Reservation.

Thunder rattles the windows. I'm startled awake in my basement bunk at four in the morning by constant flashes of lightning illuminating the room at strobe-like intervals. The thunder's vibrations roll down the stairs. I anticipate the sound of rain but it never comes. Directly above me, the phone rings several times on the main floor, followed by voices, hasty footsteps, and the front door slamming. This can't be good.

It's been a year of drought in North Dakota. Normally the Froelichs put up 2,500 enormous round bales of good hay for the winter, but this year the grass was so short it couldn't be cut. That's when it's time for a rancher to choose: sell off cattle quickly when everyone else is selling, or buy feed when everyone else is buying. Either way, the rancher loses, even before taking into account that a drought year means the threat of range fires. Dry grasses plus dry lightning storms equal a highly combustible duo.

At breakfast, Kathy Froelich fills me in on the details. The early call to her husband, Rod, was from the fire chief at the Standing Rock Reservation. A lightning strike on the edge of the ranch sent grassfire blazing across the property. The key to containing a range fire is an instantaneous response. Every rancher in this area has a pickup with a pump and a large water tank in his truck bed. This morning, the quick action of the local Native American fire fighters, Rod, and his neighbors kept the fire damage to fewer than a hundred acres. But there's still the threat of more lightning—a threat to more than a century of ranching here in south central North Dakota.

Rod's German grandfather, Matt Froelich, was just nineteen years old when his family emigrated to North Dakota from Russia. Catherine the Great had promised German farmers wealth and privilege if they would come and cultivate farmland along the lush Volga River valley. But the promises proved false, and in the early years of the twentieth century, thousands of German families in Russia risked everything to move to America.

Matt first worked for the Northern Pacific Railway and eventually saved enough money to buy foreclosure lands from a local bank in 1928. The

Froelich Ranch was finally established, though not along Russia's Volga River as Matt and his family had originally envisioned. In 1948, the ranch began to breed quarter horses when Matt's son, John J. Froelich, bought a registered Palomino stallion named WR's Commanche Boy. John's love for the Palomino breed has been passed down to his five sons, whose families abide by a governing philosophy to breed horses that they themselves love to ride. The quarter horses of Froelich Ranch play a huge part in daily operations, from herding cows to arena roping. Buyers come from throughout the Midwest to acquire these working horses with their friendly dispositions. The Froelich

Dry lightening storms bring fire from the sky but no quenching rain.

family displays great pride in their horse bloodlines. A large quote on their quarter horse Web site insists, "Yes, there's always fence to fix and cows to check, but we always have time to chat about our horses!"

Speaking of chatting, with the wildfire under control and the ranch back to normal—if you can call it "normal" during drought conditions—we finally have time to chat about the ranch as Rod treats me to a trip to the local café. It's the kind of coffee-and-pie place where everybody knows everybody, and the customers greet Rod as a good friend. I've noticed that humor seems to be a specialty of ranch families out this way, especially when it comes to obstacles. Just as quick action is the solution to containing wildfires, humor is the answer to hardships on the ranch. So here in the diner I see a rack of fly swatters by the door. Every customer who steps through the threshold takes one. North Dakota flies, huge and unavoidable, are a real nuisance. Patrons who come in for coffee swat happily to keep their tables clear of bugs. And they laugh while they swat, great booming laughs like life's going to be all right, never mind the flies and fires.

This positive disposition is found rooted in Kathy's history as well. Her great-great-great grandfather was Chief White Shield of the Arikara tribe, a branch of the

Chief Sitting Bear, a noble and good leader of the Arikara people, was Kathy Froelich's great-grandfather.

Rod Froelich

Caddoan people. Known for their affinity for peace with the region's white settlers, the Arikara were a rare non-nomadic First Nation people. Adapting to the cold winds out of Canada, the Arikara lived in windowless sod houses. "In fact," Kathy told me, "they called the whites' sod houses with their windows 'houses with eyes.'" To supplement their farming, many Arikara worked as scouts for the U.S. Cavalry.

While Kathy holds a Ph.D. in education and teaches at a local junior college, Rod, in typical ranch humor, says he holds an OBB, an "Out Behind the Barn degree." I have to smile since Rod's statement comes from a rancher who somehow finds time to serve as a North Dakota state legislator.

At the mercy of wild elements, their fate at times decided by which turn a fickle North Dakota wind takes, the Froelich family continues the dream passed down by hard-working German farmers and the peace-loving Arikara. It's a dream undeterred even now by the low, gravelly, and familiar roll of thunder.

Kathy Froelich

Granddaughter Brenna Guenther

Lance Froelich

Daughter-in-law, Sunshine Froelich brings in the horses.

Alan Chew discovered the half-eaten carcass of a yearling cow near the bales of stacked hay that had been stored for the tough winters here in the Rocky Mountains. Something big and mean must be stalking the area. On the ground, soft dirt had captured impressions of the vicious four-legged intruder. Alan dropped to one knee and studied the tracks. Each print had four teardrop-shaped toe pads, no claw marks.

Mountain lion.

Alan stiffened and stood, eyeing the eighteen-inch aisles between the rows of hay bales. Could a lion have burrowed toward the back of the stack and decided to make a winter den in the hay? While they prefer deer, mountain lions will certainly feast on livestock and pets if given a chance. This cat, nearly invisible behind piles of hay, would simply pick his meal from among the smorgasbord of cattle and sheep wandering the ranch. Easy prey.

The rancher hurried to the house and called one of the hunting dogs to his side. As he neared the haystack this time, Alan's fingers tightened on the base of his

The Chew Ranch raises both sheep and cattle. Their headquarters is next to Dinosaur National Monument.

CHEW
RANCH

UTAH

pistol, waiting for the dog's response. Sure enough, the dog immediately alerted to the presence of not just an animal's den, but the animal itself. The mountain lion was in the stack.

Alan squeezed between the hay bales, keeping his dog on a short leash as he crept further into the maze of hay. The pistol was ready. Suddenly, there in the shadows crouched the cornered cat—about five feet in front of him. Alan fired off four rounds.

I'm holding my breath as Alan, one of the four Chew brothers, tells me the mountain lion story while we ready another truckload of sheep to be driven to market. It's late afternoon at this Colorado sheep camp, a rustic headquarters where the sheep are sorted after being brought down from the high country.

"So you got him, huh?" I say to Alan. "Now I know why you always carry a gun. I bet that's one of the most important things a cowboy needs out here."

"Actually," Alan's brother, Scott, interjects, "there are only three things every cowboy needs."

I wait for this revelation while wiping my hands and setting the camera up for shots of the sheep loading into the two trucks.

"Good boots, a good saddle, and a good bed," Scott says. "Because you'll always be in one or the other."

I can understand these cowboys needing a good bed, especially after a hard day like this. Two truckloads of sheep had already set out before I arrived. Neil and Doak are the brothers who handle the sheep while they're on the ranch, while Doak is joined by Scott to drive the livestock to market. The clock is meaningless in this operation; the trucks rumble in empty and go out packed with bleating sheep into the wee hours of the night. The brothers often keep going on as little as two hours of sleep.

"What do you guys do when you get tired?" I ask, thinking of those long, steep Rocky Mountain passes where drivers have to stay constantly alert. Is the secret lots of cowboy coffee?

The brothers look at each other. "We just trade trucks!" one of them says. Sometimes I don't know if this city slicker is being taken in or not. I just smile and create a few more pictures amid the bleating of sheep.

The Chew Ranch headquarters, straddling Utah's Green River near the Dinosaur National Monument, found its humble beginnings in a Mormon caravan. In 1865, thirteen-year-old Jack Chew and his family joined 29 other families in a journey from Iowa to Salt Lake City. These Mormon travelers all pulled handcarts containing provisions and belongings. The carts had solid wheels and were two to three times the size of a large wheelbarrow.

The story goes that Jack fell in love with a young lady whose brothers had promised her to a polygamist. However, Jack secretly married her, angering the

The old homestead located far into the back country is still used.

Alan Chew moving cattle along the Utah Colorado border.

One of the boys jumps right into his work.

Counting sheep, but not to fall asleep.

The directions to the sheep operations, are to turn left when you get to the Clark Store.

brothers so much that they plotted to kidnap the new bride. They decided to frame Jack for stealing a cow—which he didn't do. He was arrested and thrown in jail. His sentence was eighteen months in prison or a fine of $25. Jack told the judge he'd need to be out of jail for one day to get his hands on the money. The judge agreed. As soon as Jack was released, he stole one of the judge's cows, had it butchered and sold in another town, and returned with the $25 to pay his fine. This quick lesson in the value of cattle may very well have been the impetus that spurred Jack Chew to become a rancher.

Jack lived to the ripe old age of 102. Doak, Alan, Scott, and Neil still tell tales about their great-grandfather. Even during the last thirty years of his life, old Jack handled the day-to-day ranch chores such as pitching hay in spite of his advanced age and blindness. He'd tie a string trail from the house to the barn, a tactile map of thread to help him get where he needed to go about the ranch. Neil tells me Jack was baptized twice in the Mormon faith, but that "neither took." I'm not sure if this is another joke.

Scott's the cowboy-philosopher of the bunch. Living near the Dinosaur National Monument, the boys as they grew up were constantly reminded of what causes the extinction of things—such as the slow disappearance of the great family ranches of the West. "The dinosaurs roamed for 280 million years, but something happened, and they couldn't evolve; they couldn't change," Scott says. "In order to stay in the ranching business, you gotta be able to change with the times."

And the Chew brothers are flexible about changing. For example, they're adding new revenue streams such as tourism and hunting to their enterprise. While the cattle business is headquartered in Utah, the sheep-raising operations

The main ranch is located on the Green River near the borders of Utah, Colorado and Wyoming.

When it's time to ship
sheep, everyone pitches
in — family, friends
and neighbors — even
the littlest cowboy.

are located just north of Steamboat Springs, one of Colorado's most popular ski resorts. Jet-setters land their private planes in Steamboat, but the Chew boys tell me it's still a cowboy town. While I'm visiting, two Texans have flown in to hunt for elk at one of the ranch properties near the Utah border. Another group has just returned from a successful hunt.

It's quiet now as the last truck rattles away. The bleating of sheep fades in the night. Steamboat Springs' lights are a glow on the far horizon, but here at the Chew Ranch the sky lights up with innumerable brilliant stars seldom seen by most Americans. My muscles ache as I pack away the camera equipment. My feet hurt, and I'm bone tired. Scott is right: there's not much I wouldn't give right now for a pair of good boots, a good saddle, and a good bed!

Helen Chenoweth-Hage in the ranch pantry.

Wayne Hage with two long time family friends.

I've been driving all day through a smoky haze caused by California wildfires. The sky looks like a reflection of the gray-brown desert, engulfing all of Nevada. I turn off pavement onto a gravel road, and the car kicks up dust that only adds to the eerie evening air. Phone calls to Pine Creek today went unanswered, so I don't have detailed directions to the ranch headquarters in one of the most remote places in the contiguous 48 states. Gravel gives way to dirt. Rutted paths, scratched deep into the landscape, veer off at various intervals like lines of a giant maze. At times it's hard to tell if I'm even still on the main road. When dusk deepens around me, the blur between horizon and desert adds to my disorientation.

Thirty miles later, my headlights illuminate a closed gate bearing the sign "Ranchero Drive." No mention of Pine Creek Ranch. I decide I must be lost but get out anyway to try my luck at the gate. It's locked. Though unsettled at the thought of spending the night in the middle of the black desert, I've been blessed with an adaptive ability to fall asleep almost anywhere, so I climb back into the car and get comfortable. Maybe in the morning I'll double back and follow another of the roads.

Sleep settles over me as I contemplate my research so far on Pine Creek Ranch, and those who lived and died for it.

In the 1860s, stagecoach drivers would call over the noise of the thumping wheels, "Next stop, Widow Smith's!" Mr. Smith had been bushwacked because of an argument over a yoke of oxen, leaving behind a grieving widow and the family land. Precedence demanded she sell and move to town, but the Widow Smith clung to the ranch house with everything she had. "One tough lady," the driver would say, readying the stage for the next leg of the journey.

Decades later, several days' ride north of Widow Smith's place, 14-year-old Wayne Hage was fed up with mandatory church attendance enforced by his mother. One night he finally spelled out his good-byes on a piece of paper and ran away from home, dreaming of an adventurous cowboy life. He didn't run far, however, and soon signed on as a cowhand at his uncle's ranch. Within a few years

Wayne Jr., Yelena and Bryan Hage.

PINE CREEK RANCH

NEVADA

Wayne Hage riding back from checking on some cattle.

Wayne became a buckaroo at the PX Ranch, the Nevada spread that Bing Crosby bought so his three boys would learn the value of hard work far from the Hollywood lights. Here, Wayne honed his skills before joining the Air Force. After a tour of duty, he enrolled in college and graduate school at the University of Nevada-Reno. With a masters degree in organic chemistry, Wayne finally returned to ranching.

In 1978, Wayne purchased the Pine Creek Ranch—over 1,000 square miles of dirt, streams, cattle, pastureland, and sagebrush that included old Widow Smith's place. He moved to the ranch with his wife, Jean, and their five children. Wayne was living his dream, but he, like Widow Smith before him, was about to discover the heartache and struggle involved in loving this Nevada land so deeply.

Suddenly I jerk awake. A cowboy is tapping at the car window. Behind the man, silhouetted in the bright interior of a truck cab, a young woman and baby watch as I sit up, fumble with the door latch, and set off the car alarm. The noise and lights pierce the cool night air. "Are you Wayne, Jr.?" I ask, finally silencing the alarm and grabbing my shoes.

His face breaks into a smile. "Yes, sorry! We're late getting back from town." We decide I should follow their truck to the ranch house, where the Hages make me feel right at home. I'm thankful for a soft, comfortable bed after sleeping behind the wheel. Morning brings the smell of fresh coffee. It's not even 6 a.m., but Wayne Jr. has finished the early ranch chores and is back inside for breakfast. Between bites of sausage and egg, I ask questions about the ranch's history.

A year after Wayne Sr. purchased Pine Creek Ranch there was a knock at the door—the U.S. Forest Service was calling to try to purchase the ranch and the water rights. The price they offered was half of what the Hages had just paid for it. Wayne's refusal to sell brought the full forces of the government's eminent-domain policies against Pine Creek Ranch. But the bureaucrats had squared off with one tough cowboy. Years of legal battles raged, and Wayne Hage's meticulous research on the rights of property owners resulted in the unsettling account of the fight in his 1989 book, *Storm Over Rangelands*. During the 90s, Wayne Sr. became one of the country's leading advocates for private property rights.

The physical and emotional impact on the Hage family, however was immense. Some family members believe the stress contributed to Jean's heart attack and early death.

Wayne Jr. offers me more coffee, and I accept as he continues with his story. In 1999, Wayne Sr. married Helen Chenoweth, a highly regarded congresswoman from

Sunset behind the ranch headquarters.

Wayne saddling up for a day's work.

Idaho. Although he suffered from cancer, Wayne traveled the country with Helen to speak and advise on private property rights. Just two months before my visit here to Pine Creek, Wayne Hage Sr. succumbed to the disease.

Wayne Jr.'s wife, Yelena, and baby, Bryan, soon appear in the kitchen. I'm surprised to learn Yelana is from Kazakhstan. I tell her that Central Asia is one of my favorite places in the world to visit. "I've spent time on layovers in Almaty on my way to Kyrgyzstan."

Yelena is astounded that a visitor to this remote ranch in Nevada knows anything about her hometown literally halfway around the world. "My mother lives now in Almaty! In fact, today is our day to video-conference. Bryan has to look good for Grandma!"

Pine Creek Ranch is remote. The post office is an hour away, and the nearest neighbor is a half-hour up the valley. Power lines don't run this far out, so the ranch generates its own electricity. And yet how amazing that Yelena can video-conference via the Internet with her family some 13,000 miles away. She informs me that her mother-in-law, Helen, will arrive at the ranch after lunchtime as she returns from a speaking engagement in South Dakota.

Riding in the pickup with Wayne Jr. is like cramming a semester's history course into a few bumpy hours. As we check cattle and fencing, he points out landmarks and relates the stories behind them. Wayne can tell you the date a Native American homesteaded in the area, who was Nevada's

The Pine Creek Ranch fills this valley.

On a rotating basis, one year out of seven, Wayne rests the land letting the native grass thrive.

Baling wire, duct tape and mason jars are indispensable items on any ranch. You can never have too many chains in the barn.

Above and in the black hat to the left are members of the Berg family, the closest neighbors.

governor any given year, and what dates Congress passed certain laws that have affected the West. People say he is as well read and knowledgeable as his father was.

The evening meal is delightful as I meet Helen, and the conversation ranges from Central Asia to cowboy culture to life inside the Washington D.C. Beltway. I ask Helen how she and Wayne Sr. met.

"I'd been single for over twenty years and was very comfortable with that," Helen says. "My life as a congresswoman was very full, with little extra time even when back home in Idaho. Wayne's oldest daughter, Margaret, was a business friend of mine, and she wanted me to meet her dad. I told her I didn't have an interest in dating or a relationship. Later, I was talking to her prior to leaving Washington for a week at home, and mentioned how much I looked forward to resting.

"Margaret said, 'If you really want to get away from the phone and any interruptions, you should come down to the ranch. Bring a few books, because there's not much to entertain you.' So I did. I met Wayne and he tipped my canoe!

He was the most remarkable person I'd ever met, and it wasn't long until we married. My term in Congress was up, and I moved here."

Each night, Helen tells me more about Wayne, Sr. She says he'd often put a Bible in his saddlebag, reading in the shade of a tree at noon or beside a night-time campfire. In spite of wanting to put religion behind him as a boy, Wayne found answers to many of life's questions in the Bible.

Several Bibles are in the huge ranch house library, where one wall of shelves is completely stocked with law books. At supper on my last night at the ranch, I ask Helen if I can take her portrait in the library before I leave. As the shutter snaps and captures Helen the following day, I feel keenly aware of the huge contributions this family has made to ranching endeavors and land ownership rights in the West. I pack up my supplies after I've said my goodbyes and watch the ranch receding in my rearview mirror.

Nine days after my departure, Helen, Yelena, and Bryan start out on a typical 65-mile trek from the ranch to the post office, unaware of the tragedy awaiting them.

Helen Chenoweth-Hage in the ranch library nine days before she died.

There is a terrible automobile accident. Miraculously, Yelena and the baby survive, but Helen is killed instantly.

Sorrow descends over the Monitor Valley, but still the Pine Creek Ranch endures. The Hage family soldiers on, hanging in a delicate balance between desert and mountain, between civil liberty and government intrusion, between hearts broken and dreams realized. Will baby Bryan's little hands grow strong enough to grab hold of his future here?

Meg Keenan, Helen's daughter, leads Helen's horse behind the wagon to the burial site on the ranch.

Helen's body lies here beside Wayne, Sr. and his first wife, Jean, but their spirits live on in the presence of their Lord.

Meg receives the American flag from the coffin, among family and friends.

I've been lying awake for several hours. Adrenaline runs full force through my veins. Soon it will be dawn and the adventure will begin. This ranch is the first one for me to photograph in the *Great Ranches of the West* project, and being a city guy, I'm not sure what to expect. I have only ridden a horse three or four times in my adult life, and today would require me to be out on one all day.

I meet the cowboys down at the corral and see that they are each taking three horses and loading them onto trailers. The ranch foreman is leaning on the fence of a corral that's noisy with the snorts and whinnying of 50 horses. Known for its working quarter horses, the Reeves Ranch breaks and trains their animals from the very beginning to work cattle.

I ask the foreman if he'll pick out a gentle horse for me. I'm picturing an animal that's small and mild, perfect for wandering through the South Dakota countryside at nothing more than a slow trot. After a few puffs on his cigarette, the foreman squints, surveying the selection in the corral. Turning to me with arms folded across a rock-solid chest, he says, "Mr. Keen, every horse here this morning

Dean and Emma Lu Reeves

has been ridden at least once, but I see only a few that have been ridden more than three times. You think you can handle one of them?"

I nearly drop my camera bag, stammering, "Doesn't Dean Reeves or somebody have an older horse? Maybe one that's been ridden for a few years?"

The ranch foreman turns on his heel, making his way toward one of the horse trailers. He calls over his shoulder, "We've got two hundred horses to train this summer. Nobody has the luxury of riding a favorite. Every time we move cattle, each of us is training a new horse."

REEVES RANCH

SOUTH DAKOTA

He closes the door to the trailer and nods toward the barn. "But I guess you can have that one." I look in the direction of the nod, and there sits a four-wheel all-terrain vehicle. I let out a huge sigh of relief.

Exploring the Reeves Ranch on an ATV is both exhilarating and sobering. The drought in South Dakota has been severe again this year. Grass that normally boasts a foot and a half of bright green growth shows stunted, brown shoots four or five inches high. The reservoir on the Missouri River has receded drastically, leaving vast, spongy mud flats. Decaying cattle carcasses appear now and again out in the mud — a solemn vision.

I stop my four-wheeler to investigate what was formerly a lake. The mud holds firm as I begin to walk the first twenty yards, but each step past a certain point feels less stable underneath my feet. Suddenly, one foot breaks through the crust. I pull back, but not before muck oozes into my boot, and I nearly lose my balance. This is sticky stuff. I can understand how cattle seeking the water's edge become trapped. Once a cow's hoof cracks through the outer scab of mud, each struggling move sinks the animal deeper and deeper. It's not quicksand, but it acts like it. If a horse-riding cowboy happens upon the scene in time, a carefully orchestrated roping will pull the stranded cow to safety. But the carcasses dotting the landscape bear witness to the fact that, often, it's too late.

Running this first-generation ranch has been a tough go for the Reeves. Both Dean and Emma Lu grew up on ranches, so when they leased these 20,000 acres from the Standing Rock Indian Reservation in 1961, they knew what they were in for. It was virgin land, untouched by man or domesticated beast. Of course, nomadic Native Americans, French-Canadian trappers, and frontier families heading west moved through this region, but no one had ever settled here; no one had called it home. The Reeves built a two-room cabin with a coal stove, hauling water from the river for household use until 1980.

Dean established the ranch's horse-breeding standards with ten athletic quarter horses that were gifts from his father. In 2000, the American Quarter Horse Association recognized Dean with a unique award for his 50 years of

Dean and Emma Lu's three-year-old grandson.

Rounding up bucking horses for a bronc riding seminar.

Bringing cattle up away from the mud at the edge of the Missouri River.

champion quality quarter horse breeding. The patience and perseverance of the Reeves family is seen in their son Tom, who consistently placed in the World Saddle Bronc Championship but never won throughout his seventeen years of competing. Showing a spirited determination, Tom finally triumphed, becoming the 2001 world champion at the unheard-of "old" age of 36.

It takes serious resolve to make it on this unforgiving land. Central South Dakota always has severe winters, but the winter of '96-'97 proved an especially horrendous year for wind and blizzards. Mounds of snow completely buried the rear of the ranch house. After one particular freezing storm finally let up a little, Dean fought the front door open and took a step out, intending to go check on the cattle. He wasn't sure who had the greater shock: he or the huge buck lying in the shelter of the front porch.

Later that year, an odd April warm front with temperatures in the eighties melted the fifteen-foot-deep snowdrifts into streams and pools of water and mud. The cattle naturally waded into the streams for water. The warm temperature was quickly followed by a cold northerner, and they waded farther into muddy pools to escape the wind. That night the temperature plummeted far below freezing, and a full blizzard set in. By the time the weather broke and the ranchhands could get out, hundreds of cattle were dead or dying, trapped in the frozen mud. The worst part of the horrific experience, says Dean, was having to don wet suits and haul out the dead animals one at a time. The rotting carcasses had to be buried quickly nearby. It was weeks of morbid work, and the Reeves lost more than 200 head of cattle.

ARTHUR DEAN REEVES

AQHA
AMERICAN QUARTER HORSE ASSOCIATION

50 Cumulative Years of Breeding
American Quarter Horses
1943–2000

But the Reeves are ranchers. Emma Lu follows a stringent philosophy of always emphasizing the positive. In this demanding landscape, you need to find the beauty where you can, and you need all the encouragement you can get. That's just the South Dakota way.

It's late at night and I still have three or more hours of driving before reaching the Siddoway Sheep Ranch. I haven't seen another car for some time now, and I'm road weary. Rounding a bend in the road, I notice light coming from a lone motel in a small Idaho town. It's either stop here for the night or risk falling asleep behind the wheel and ending up in a cold ditch somewhere, so I pull in.

I'm greeted in the lobby by the night manager. His face is weather-beaten. Muscled shoulders and arms bulge under his shirt. He doesn't have a hat on, but I bet this guy is a buckaroo moonlighting to make ends meet. His ramrod straight back is a sure sign.

Though my camera equipment is most likely safer here in rural Idaho than in any city, I don't feel comfortable abandoning several thousand dollars of gear in the car. I carry the equipment inside the motel room, leaving my clothes in the trunk because I'm too tired to make yet another trip back out to get anything else. I hit the sack for a few hours of sleep. Come morning, I wander into the lobby for a cup of coffee. Whoa! Whoever made this brew must have an attitude, because the coffee's thick enough to float a horseshoe. I am now convinced that this motel manager—maybe he's even the

The littlest shepherd, five year old Wayne Siddoway-Richey does his part. Most of the sheepherders on the Siddoway ranch come from South America.

SIDDOWAY SHEEP RANCH

IDAHO

owner—is a former rancher. Like many of his contemporaries, he's been through financial storms, maybe lost his ranch. This coffee is his way at getting back at the world.

In a caffeine-fueled burst of energy, I head for the Siddoway Sheep Ranch. The Siddoways are herding their sheep down from the summer pasture in the rugged Teton Range above Jackson Hole, Wyoming. The family and their sheep have retraced this journey on the border of Wyoming and Idaho since first settling here in 1886. The undertaking, then as now, is not without its dangers. A few nights ago, before the herders began bringing the sheep down to the shipping point, a wolf sneaked past the guard dog and killed thirteen sheep. The quiet attacker didn't eat any of them, but killed the sheep for the sheer fun of it before being driven away. The wolves that were reintroduced into Yellowstone National Park have rapidly bred and expanded into many packs that have wandered far from the Park boundaries. These killing machines cause great financial loss to ranchers.

More loss has come from the depressed wool and lamb market now that cheaper, imported wool has gained such a foothold. Jeff and Cindy Siddoway have been forced to diversify their operation in ways never dreamed of by their ancestors. Inspired by an episode of *Little House on the Prairie*, the family was determined to make something good rise out of a hopeless situation, and made a bold move into the world of quality wool blankets. Siddoway blankets feature ranch-themed jacquard wool throws and quality seasonal trade blankets, using colors and scenes inspired by the surrounding Teton Peaks. The Siddoway Wool Company, LLC, even has an engaging Web presence for marketing purposes, offering to ship "anywhere." Branching out even further, the family has fenced off 11,000 acres of their land to create a hunting lodge and preserve. Guided hunts for trophy elk and bison are available to guests. By broadening their horizons, the Siddoway family has ensured success for future generations, no matter which direction the economy takes.

Truth is, these folks have always been resourceful, a gift passed down from frugal and determined pioneers.

Jeff Siddoway

A traffic jam on a back-country road.

Jeff's great-grandparents James William and Ruth settled in Teton City in 1886, living in a tent while they built a home out of logs. Transplants from Salt Lake City, the Siddoways began a legacy when they registered their sheep brand in 1898. Their descendents include highly-educated men and women well-versed in animal husbandry, business, and law, not to mention a long-standing and deep passion for sheep ranching in these picturesque mountain ranges. While I'm visiting the ranch, Jeff is running for a seat on the state senate for his district of Idaho. He would go on to win in the November elections, a natural progression for a man who's already president of the Fremont Wool Growers Association, past president of the Idaho Wool Growers Association, and a past member of the Idaho Fish and Game Commission.

Diversification has been difficult but necessary in these days of uncertainty.

Moving the sheep down from the high country.

Cindy Siddoway was the first woman to be president of the American Sheep Industry Association. She served from 1999 to 2001. She and her son J. C. are taking a short break from loading sheep on trucks on a cold fall morning.

Back at their house that evening, we all thawed out as the rains came down.

In order to keep the Siddoway dream alive, family members have sacrificed time and effort, many of them splitting hours between a "real" job and sheep ranching. It's not a profession just anyone could assume.

Jeff's father Bill found this verse carved into an aspen tree on the property. The author remains unknown, but the sentiment is good-naturedly understood by the Siddoways and every sheep rancher who is determined to carry on in the face of adversity:

> I've summered in the tropics,
> Had the yellow fever chill.
> I've wintered in the Arctic,
> Known every ache and ill.
> Been shanghaied on a whaler,
> And stranded in the deep.
> But I didn't know what misery was
> Until I started herding sheep.

The hunting lodge at Juniper Mountain Ranch.

Elk and buffalo are some of the animals that roam Juniper Mountain Ranch, the Siddoways 11,000 fenced acre, hunting ranch.

The ship's cabins didn't make a good first impression. The ladies stood in one cabin's doorway, peering into the small, dank cavern as their husbands up on deck were being briefed by the crew on the ship rules — rules that this captain strictly enforced. The men then received their work assignments. Every passenger would have several turns at scrubbing the deck and cleaning below.

Having crossed the Atlantic twice now, Svan Anderson knew the drill. He dreaded the thought of telling his two brothers and their wives that they might have to spend a full ten to fourteen days in the same set of unwashed clothes since no washing or hanging of clothes below-deck was allowed. They might be able to do laundry once a week on deck if the weather permitted. And the North Atlantic weather was known for not permitting much.

True to form, a violent storm smashed into the ship after being at sea just a few days. All below deck windows were ordered shut to keep seawater from pouring in. A third of the passengers became violently seasick. And the following day, most of the remaining passengers were nauseated from the terrible smell. Very few showed up at the two stoves to cook breakfast that next morning.

On the fourth day the storm finally abated. The captain would hate to add to his official log a report of a cholera, measles or typhoid outbreak—all very common on trans-Atlantic crossings. So the Andersons and everyone else aboard were ordered below decks to scour every inch of the vessel to prevent a disease.

On Sunday the captain hosted a party and requested all who could play an instrument to attend a short rehearsal. The Anderson couples enjoyed the merry concert at sea that night, with dancing and singing on this clear, calm evening in 1868. As the journey neared its end, the ship encountered dense

Three generations of Anderson guys.

In a normal year, the Flint Hills of Kansas receive 36 inches of rain. Steve Anderson holds a nine-foot-tall blade of grass that he has mounted on to a stick.

ANDERSON
RANCH

KANSAS ✳

fog off the coast of Newfoundland. The captain called all the musicians back on deck and asked them to play their instruments as loud as they could so that other passing ships could hear them coming.

After landing in America, the three Anderson couples persevered through the immigration process and then the train trip to Kansas City; both were a breeze compared to their North Atlantic crossing. The land that Svan Anderson had homesteaded a year earlier was southwest of Kansas City in the Flint Hills region. There were no wagons traveling that direction, so the three couples walked one hundred miles out of Kansas City, carrying their few belongings. When they finally arrived to walk on the ground of their homestead, Svan and his family held a little service thanking God for bringing them to America.

One hundred and thirty-nine years later, I pull onto Poor Farm Road, wondering about such a name. Steve Anderson, great-grandson of Svan, meets me at the door of the old rock house, the home his ancestors built in 1885.

My first question is about the name of the road. Steve smiles and tells me the story. "In the days before the state had a social welfare program, my mother and father—Oliver and Mini—took in poor folks who were down on their luck. They would stay here at the poor farm until they could find work and move out on their own. That's one of the big lessons my father taught me. Show kindness to everyone, especially the underdogs."

Oliver knew firsthand what it was like to be poor. During the Great Depression he hopped a train looking for work. At every stop, a couple hundred men would get off and be replaced by another couple hundred clambering on. Oliver had heard that he could find work in Colorado picking peaches. When he arrived at the orchard, he hadn't eaten for days and had no money. The boss there said, "Sorry, Bud, but I've already hired the workers I need. And there are a hundred names on the list ahead of you. Good luck." And he tossed Oliver a dime.

"One dime," Oliver thought as he picked it up. "What good will that do me?" He stuck the dime in his pocket and walked into town to the general store, wondering how best to spend it. He settled on two small packages of biscuits for his lunch and sat down in front of the store to eat. A hollow-cheeked teenager wandered by and stared at the biscuits. Oliver sighed and offered the second packet to the boy. They sat and shared stories and parted friends.

That night Oliver joined several hundred other desperate men in the city park. They'd all be sleeping under the stars, some with an extra coat, others with just a few sheets of newspaper for a blanket. Darkness

The fire that destroyed the hay barn of Matthew Anderson and the replacement barn below.

Matthew untying a bale of hay.

descended, and the men settled into an uncomfortable sleep.

The silence was broken by someone walking around calling, "Anderson! Anderson!" Oliver recognized the voice of his young lunchmate.

"Mr. Anderson!" the lad exclaimed. "I found work today splitting firewood. The man paid me fifty cents. Let's go eat." They walked to the café down the street, and both feasted on a hamburger and a cup of coffee. The story comes down through the Anderson generations that, with a full belly and the knowledge that good deeds are

The ranch house on Poor Farm Road was built in 1885.

rewarded, Oliver faced his future with hope.

For the next few years of the Depression, Oliver worked as a laborer in the wheat fields, starting in Texas and following the harvest on up to Canada. He returned when he could to his family on the little farm in Kansas. Oliver was a big, stocky six-footer. The farmers would feel his vise-like handshake and know they could get the work of one-and-a-half men out of him.

In the 1940s, strong-man contests were part of many of the traveling circuses. Professional wrestlers would challenge local men. The professionals were taught to lose the first match or two to hustle higher bets in later rounds. One year, Oliver's friends put him up as their best bet to challenge the professional. The circus barkers, skilled at working the crowd into a frenzy, encouraged bigger and bigger bets. They expected their champion to win since he'd drawn loads of money defeating the best local men in one rural town after another. But as an untrained underdog, Oliver entered the ring and quickly ended the contest by breaking his opponent's arm!

I follow Steve outside, and he tells me about the unusual height of the grass on the ranch. When the Flint Hills receive normal rainfall, the grass can grow well over your head. I'm figuring this is a bit of a tall tale until he grabs a long 1x2 board with a blade of dried grass stapled to it. The grass measures exactly nine feet. Steve tells me, "My kids always hated to have to walk out for the horses. The dew on the grass would soak them from head to toe."

Steve was the first in the Anderson clan to expand the farm into an extensive ranching operation. His expertise in both farming and ranching was rewarded as he served Kansas Governor Joan Finney from 1991 to 1995 in the role of Agriculture Liaison. Steve recently turned over most ranch operations to his second son, Matthew. They run cattle today on their own land as well as on leased land from 25 of their neighbors.

When I meet Matt, he shows me one of their new hay barns. The old one caught fire earlier this year. The school bus driver saw the fire and radioed for the fire department. But by the time Matt got there, he was only able to save a few pieces of equipment. The barn and hay were a total loss. Matt's insurance for replacement of the hay was at last year's normal hay price. This year, however, a shortage has hiked the price of hay three times

Big Mac, the 4H steer that Matthew and Julia's son raised.

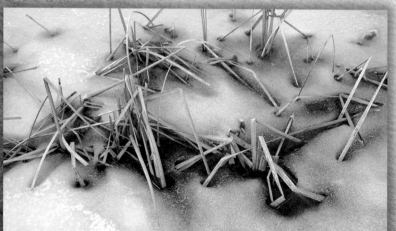

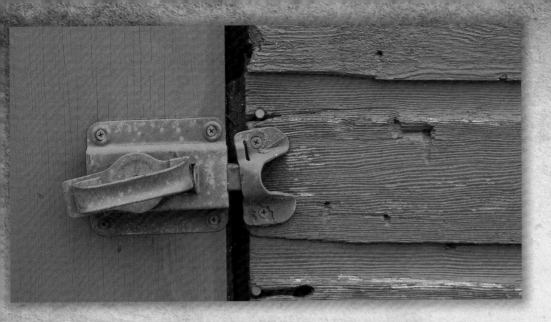

what the insurance would pay. It was a huge loss. But here's where the heritage of the Anderson family's kindness principle brings the return on investment: Used telephone poles were donated, used roof trusses were offered cheap, and then neighbors and friends gathered to do the work. And there on the Kansas plain well more than a century after Svan plowed his first furrow, the Andersons and their friends celebrated an old-fashioned barn-raising.

As I gather up my equipment to drive toward Colorado and home, Steve shares another Anderson principle from his tough old father Oliver: operate your business with no debt. When hard times come, if you have to pay your banker first, there may not be anything left for the family. With typical cowboy humor, Steve has written a "Profit Tips" poem—verses of which echo in my mind as I turn off Poor Farm Road and head west:

> Sell your cows, sell your sows,
> There's no profit in pulling plows.
> Quit baling hay, take time to play,
> There's no profit working day after day.
> There's no profit in chicken, no profit in beans,
> Quit while you still got a nickel in your threadbare blue jeans.

Matthew and Julia Anderson. He runs the ranch full time now that his dad has retired. He also serves on the board for the Lutheran School in the nearby town of Alma.

The grave of Gottfied Nehring, the first from Steve Anderson's mother's side of the family to settle in the area.

One of Fred Eiguren's hired hands found this old gun buried in the sand on the ranch.

Richard Eiguren closing the gate at the end of the day.

Sheep rubbing against the fence leave some of their wool on the barbs.

The sagebrush was huge, sometimes towering over them and their shared horse. The brush was so thick that a straight path wasn't possible. The frequent backtracking slowed their progress and used up their slim resources. Jose Navarro and Antonio Azcuenaga began their journey from Nevada in 1889 to seek their fortunes farther north. Originally from the Basque country in the western Pyrenees Mountains of northern Spain, they found this territory very different and difficult. At the beginning of their trek, the Nevada desert was barren, but at least by traveling in the cool mornings they could make good headway. Now, on the high desert of southeastern Oregon, they ran out of food and water. And the thick sage made travel exhausting. Jose and Antonio were both close to death when a trapper stumbled upon them. He gave them provisions and directed them to a route where they would find water.

Jose eventually settled near Jordan Valley and started a sheep ranch. He returned to Spain to marry in 1900, and a year later his twelve-year-old nephew, Pascual Eiguren, came to Oregon to work for him. In a few years, Pascual became a full partner with his uncle and started his own ranch. By the 1920s, eighty percent of the people in this corner of southeastern Oregon were Basque.

In those early days, there were no banks in the region. Pascual took the initiative to finance the projects of many of his Basque countrymen. He was a conservative, benevolent man who did not charge interest. Once, while taking his sheep to market in Idaho, some of the Idaho ranchers were belittling him because his sheep were smaller than theirs. Another Basque shepherd, one Pascual had helped financially, cut the joking short when he pointed out that Pascual owned his sheep outright—while the Idaho ranchers owed so much money to their big city bankers that their sheep could hardly be called their own. Soon the Great Depression of the 1930s hit, and many of the Idaho ranchers were out of business, along with their banks.

FIGUREN RANCH

OREGON

Every granddaughter should have the right cowgirl "stuff" including pink boots.

Today Richard Eiguren takes me on a tour of the Eiguren cattle ranch in a "mechanized mule." It's like a small open-air pickup, just a little bigger than an ATV. I ask, "When you bring the cattle into the corrals for branding, do you rope them or run them through a chute?"

"Through a chute!" Richard says. "Are you kidding? You might as well work in a gas station if you are going to be that kind of cowboy. Guys who do that don't even know how to rope!"

I'm reminded again of something all ranchers seem to have in common: strong opinions.

Geese, ducks, and other birds fill the streams and small ponds that we drive by. Passing a wheat field, we startle some pheasants on one side of the trail and sage grouse on the other. Richard prefers to use this "mule" on many parts of the ranch since it is gentler on the land than his pickup. When I remark about the abundant wildlife, Richard mentions the Malheur Migratory Bird Refuge nearby, established in 1908 by President Theodore Roosevelt. Richard says, "We have just as many birds here as they have. In fact even the bugs prefer our ranch over the bird refuge."

We are driving several miles towards the home of his brother, Fred. Their father split the ranch in half for his two boys, and Fred lives on the original homestead. The Sheep Ranch Fortified House sits on Fred's property. It's the family's former dwelling, a historic rock structure built by the U.S. Cavalry during the 1860-era Indian wars. Although there were nearly 8,000 Paiute and Western Shoshoni Indians in the region, only about 1,100 lived on reservations. The rest were still nomadic, with renegades among them raiding settlements in Idaho, eastern Oregon, and northern Nevada. Indian warriors and Cavalry troops fought back and forth across this area for many years before the Eigurens settled here.

One day as a ranch hand rode through the sagebrush, he spied the butt of a rifle sticking out of the sand. It most likely belonged to an Indian and had been in the sand for nearly a hundred years. It was in surprisingly good condition due to the high desert's arid climate.

We arrive at Fred's house, but he's not in. Since he's the family historian, I make plans with his wife Kristie to come back the next day. I'll be driving the mule in the morning to reach a perfect spot for a panoramic image of sunrise over the ranch. So Richard says to just head on over to his brother's from there. "No problem, I'll see you at noon," I reply.

The next morning, it's still dark when I head out the door. The sunrise is stunning, and after catching the right photos I wander along, driving the mule, stopping at ponds and anything else that catches my attention in the early morning light. Soon it is mid-morning, and I figure I'd better head to Fred's place. As I drive back, the abundance of side trails shooting off my route begins to look unfamiliar. You guessed it: I've made a wrong turn. But I feel confident I'm heading in the right general

Lower right: This young cowpoke surprised himself with a bucket of cold water, or did the horse tip it on him?

direction. As soon as I get to that ridge I'll see Fred's house in the next valley. I rev the mule up the steep rise, and—.

There is no house in the valley. Maybe I'm on the wrong ridge. I aim a little farther west on the ridge and then descend into the valley. Bouncing in my mule across the basin floor, I come to a large stream, but the banks are too steep to cross. Now what? An hour later, I find myself back where I started. Let's just say that the experience gives me a vivid perspective on Jose's journey more than a century ago. Every ridge and valley looks identical. Without a map, you're lost.

Fred takes me over to meet his dad Fred Sr. His hands are so weathered and rough that shaking hands with him is like grasping a wood rasp. When he grew up on the ranch there was no electricity or running water. There was no plumbing either, so you used the outhouse even at thirty below zero. The ranch work was all manual labor; the Eigurens didn't even get a tractor until 1937. Because harvesting was done by hand, the family would have to hire fifty men just to help with the haying in those days. You had to be tough.

And you still have to be tough. It's in the air. Richard and Margene Eiguren's three sons are all tough cowboys who rodeo every chance they get. But they know it's not just physical toughness ranchers need today.

Elias Eiguren is studying at Oregon State University on a number of scholarships including one from FFA.

Fred Eiguren

To the right they are loading hay into a grinder that spits out feed for the calves, as you see on the immediate left.

Richard and Margene managed to get their whole family on horses for a portrait.
I'll bet that's something you couldn't do in your backyard.

Richard drives a low impact "mule" through a ranch stream.

It's also the stamina of character—which seems to come naturally to the family through the generations of hardy Basques.

This evening Margene has planned for me to take a portrait of the Eiguren generations. There are enough saddles for everyone, even the grandkids. So the entire family mounts up, and once the horses still their hooves, I take their portrait.

Pascual would be proud.

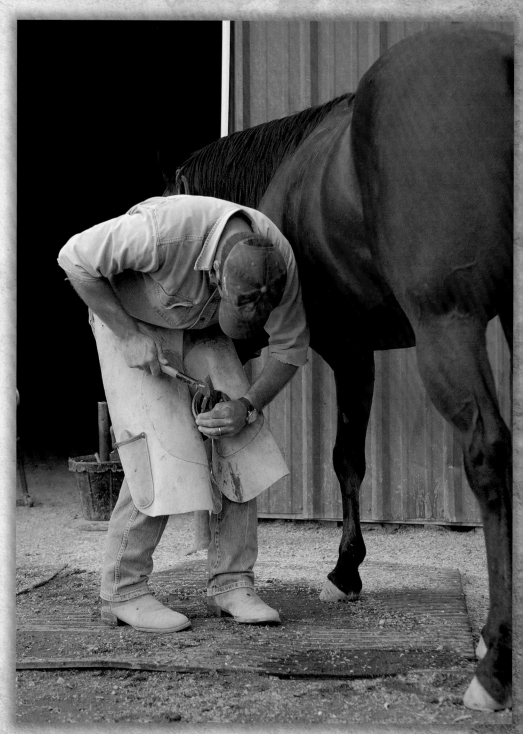

Four generations of ranch ladies wear many hats on the ranch.

One of the boys shoeing his horse.

An old rock corral and the fortified house that the US Cavalry built in the 1860s.

Richard and Margene's family room is often filled with the sounds of their grandchildren and other family members.

An old stove that still receives almost daily use in one of the outlying houses on the ranch.

Take me out to the ballgame,
Take me out with the crowd.
Buy me some peanuts and Cracker Jacks,
I don't care if I never get back...."

In the late 1930s, three brothers in Seattle, Washington were baseball-crazy. Each spring they would grab their gloves, a ball, and bat and head for a diamond singing those words. But these McIrvin boys were not your ordinary sandlot players. They were good—especially the youngest, Calvin. As only a junior in high school, Calvin McIrvin was drafted to play for the Philadelphia Athletics. He had a blazing fastball and excellent control. But no sooner did Calvin start his professional baseball career than the U.S. Army began drafting as well, and he found himself in the thick of World War II.

During the war, Calvin's father Harry bought a ranch in the beautiful mountainous country of northeast

Clive and Ruby McIrvin wearing the only clothes they have, just a few days after their house burned to the ground.

Right: loading cattle for the move to the Southern Washington winter grazing along the Columbia River.

DIAMOND M
RANCH

WASHINGTON

Washington along the Canadian border. Because of the family's love for baseball, they named the ranch "The Diamond M." Harry built a provincial, one-room log cabin that had no running water. In the winter they melted snow—which was always plentiful. In the summer they carried water from a nearby creek. Back then there were only six telephones in a 30-mile radius, and all the phones would ring at once. You had to answer, but then hang up if the call was for one of your neighbors. Electricity wouldn't arrive in the area until 1951.

After the war, Bob and Clive McIrvin traded their gloves and bats for ropes and chaps, joining their father to work the ranch. Calvin resumed his baseball career for several years, but the typhoid fever he had contracted while fighting overseas kept him from regaining his former pitching edge. Calvin didn't have an interest in ranching, so when he retired from baseball he moved to Portland, Oregon, and started a small company there.

Clive taught his son Len the ranching business. Len, along with most ranchers, has strong opinions about the government's idea to implant cows with electronic chips for identification. A few ranches have tried the system, but Len thinks permanently tracking his thousands of cattle can best be accomplished by branding his herd with the Diamond M. Branding has been a successful component of ranching since first developed in the 1500s in Mexico. Len sees no reason to mess with a good thing, especially when the new thing is costly and prone to false readings. It may work if you have just fifteen or twenty cows, but when you have thousands of cattle and have to enter each one in a computer after each relocation, it's a huge burden and expense.

As the McIrvin family grew, the ranch grew as well. Through the years they added several properties farther south. Each fall the family fires up five huge cattle trucks and transports about 5,000 head of cattle to winter in the milder weather of the Columbia River Basin in the southeast corner of Washington.

Running cattle either along the Canadian border or along the Columbia River demands more than a well-maintained fleet of trucks and an easily-identifiable brand. In the northern property of the Diamond M, the enemies are the frigid winters and mountain lions. Bill, Len's son who now runs the ranch, once bagged a lion that stretched almost nine feet from nose to tail. This lion was a serial killer, having hunted the livestock for years. It's now mounted on the den wall of the ranch house. In the southern properties, the weather is less of a threat to the cattle, but the predators are more relentless. Packs of coyotes have killed so many cattle in the Columbia River Basin that the McIrvins estimate a loss of $30,000 annually over the past five years due to coyote attacks.

I'm blithely driving along a beautiful stretch of the Columbia River to photograph the McIrvins' southern operations when, several hundred miles to the north, a Diamond M house has caught fire. Two cowboys shoeing horses just outside the barn first notice smoke pouring out of the house that Clive McIrvin and his bride Ruby moved into over a half-century ago. One hand grabs a hose and another telephones Bill that his grandparents' house is on fire.

The cowboy with the cell phone runs into the smoke-filled house and bolts upstairs, touching the scalding doorknob at the top of the stairs. It's so hot it burns his hand. He tells Bill over the phone, "If your grandparents were in here, I'm afraid they've already perished."

Four days after the fire, the remains of Clive and Ruby's house still smolders.

Bill is shaken. "Well, if you have time, toss anything valuable out the window. Be there in a couple minutes."

When he arrives, Bill sees a huge pile of deer heads and trophy antlers that have been ripped from the living room walls. It was what the ranch hand thought most valuable, of course.

Frantically searching for his grandparents downstairs, Bill phones his dad Len, who is hauling a truckload of cattle to the southern ranch. Len squeals the big rig to a

Right: The tamarack tree belongs to the larch family and although it looks like an evergreen, it does change color in the fall. The northern ranch is in this kind of rugged country.

The saddle Len McIrvin is in most often is in one of his trucks.

Bill and Roberta McIrvin, with their two daughters, live on the northern-most ranch in this book – just a few hundred yards from Canada.

stop and heads back to the ranch, an hour away. He tells his son, "Get the photographs and documents in the bedroom file cabinet. And get Grandpa's gun. And his Bible."

Bill and a ranch hand rush into the downstairs bedroom. Clive and Ruby are not found anywhere. They throw the real valuables out a broken window and finally, when the heat becomes unbearable, dive out themselves.

I arrive at the northern ranch headquarters four days later to see a six-foot-high pile of still-smoldering ashes. The house burned to the ground. Fortunately, Clive and Ruby had driven off to town that morning for a visit to the doctor and weren't aware of a thing until Bill's wife, Roberta, met them on the road home with the sad news. They'd lost everything except for a few photographs, and some deer head trophies.

A baseball or two also survive the disaster, and when the extended McIrvin family gathers for holidays, the sport is still a favorite pastime. These days, it's usually a very competitive game of softball instead of baseball, but the clan still gives heart and soul out on the diamond. Root, root, root for the home team!

The killing days for this lion are finally over.

The cattle spend the winter grazing on the wheat harvest aftermath.
Hired hands put up temporary electric fences around these pivot fields.

Bill's sister, Shirley, and her family live on this ranch shown here to the right. It is located between the northern and southern ranches.

Looking into Canada at Christina Lake, from the northern ranch.

Sam and Rosalee Britt have been married for more than half a century and have outlasted the many storms that ranching can bring.

He looked like the famed "Marlboro Man." Working his horse hard one day in the gusty ranges of northeastern New Mexico, Sam needed a smoke break. He swung off his horse and squatted with his back to the strong wind to shelter his match, reins in his gloved hand. Before he could light up, something spooked his horse, and instantly a white-hot pain shot up Sam's back. His skin reacted like he'd been seared by a branding iron. As he stood to control his frantic horse, a huge diamondback rattlesnake slithered back into the brush. Sam fingered his back, just above his belt: fang marks.

He leaped on the horse and galloped toward the ranch house, nausea surging as every part of his body swelled with poison. The horse knew the way home, and by the time they arrived at the ranch Sam had collapsed, bloated and unconscious.

Today, Sam Britt says everything good that's ever happened on the Pasamonte Ranch is the result of prayer. His wife Rosalee is known and respected for her hard work and her devotion to prayer. Over their half-century of marriage, Rosalee has prayed for the children's health and for rain. For the grandchildren's education and for rain. Safety for the hired hands, healing for hurting people in the community, and…for rain.

The church in Clayton, New Mexico where Sam and Rosalee worship.

Sunset from Jack's Point, the highest spot on the Pasamonte Ranch.

PASAMONTE RANCH

NEW MEXICO

Ranching is totally dependent on the weather. In northeastern New Mexico, even the miles and miles of famous, high-nutrient grama grass desperately need water. In 1962 it didn't rain at all. Sam had to cut down the size of the herd. Again in 1963 it didn't rain, and the Britts sold even more cattle. 1964: same story. Sam and Rosalee realized they would have to sell all their remaining cattle. "We had this huge ranch and wouldn't have a single cow," says Sam. "The place was so dry you couldn't run a goose on it!"

As Sam and his foreman, Carlos Leal, struggled through a massive dust storm to bring in the last of the cattle bound for Montana, Sam couldn't stop cursing the blowing dirt. Carlos shouted through the wind, "You shouldn't cuss the dust. Without it you wouldn't even have the ranch. Your grandaddy got to buy it because of the wind and dust!"

Simon Peter "Cap" Britt—Texas attorney, investor and director of the Federal Land Bank in Houston—founded the Britt Ranch just east of Amarillo in 1913. During the Great Depression and Dust Bowl years of the 1930s, Cap learned of the famous grama grass of northeastern New Mexico and purchased the Pasamonte Ranch. His son Buck and grandsons Max and Sam—actually Simon Peter Britt II— moved to the ranch in 1942. While Max and his dad began breeding and training race horses, Sam dropped out of college, threw his saddle into a westbound bus and landed a ranchhand job in Douglas, Arizona. Then, after four years in the U.S. Navy and marriage to his high school sweetheart, Rosalee, Sam headed back to the ranch to settle down. Just then, Max announced he was leaving for California to train their father's race horses.

"When my brother left," Sam recalls, "every one of the hired hands quit. Except Carlos. I can laugh about it now, but I had a big ranch to run and no help! Couldn't have kept it all together without Carlos."

It was Carlos in 1964 who chided Sam about cursing the dirt. And then, finally, it rained. In '65 the lakes filled, the range grass turned greener than anyone could remember, and the Britts began the very slow process of rebuilding a herd from scratch. Now, forty years later, the Pasamonte Ranch is one of the great ranches of the West.

I'm on the ranch one fine spring day when Sam and the boys are branding.

A new-born pronghorn.

Once the cattle are separated into corrals, the cowboys take turns roping, holding down, branding, giving shots, and castrating. Ranches use propane rigs to heat the branding irons, and Pasamonte's unit has a small grill attached to the top of it. The cowboy doing the castrating places the delicacy we know in Colorado as "Rocky Mountain Oysters" on the grill. Every once in a while he uses his knife to turn them for a perfect barbeque. When a cowboy has a minute, he'll grab a few for a quick snack. Sure enough, a cowhand soon offers me this treat hot off the grill. Noting my hesitancy, the cowboy—sensitive guy that he is—apologizes for his offer. After all, he says—amid the laughter of the whole crew—he realizes he's forgotten to wash his hands.

That evening Sam takes me up to Jack's Point, the highest spot on the ranch, to compose a panoramic photograph of an approaching bank of clouds. I can almost see the big stone ranch house miles away—originally an overnight stage stop on the old Tascosa-Taos Trail. I can imagine how difficult it was for travelers along this early route west.

It's here that the vast grasslands with their dry pastels give way to the sharp canyons cut by Tamperos Creek. These unexpected ravines are edged with tall pines and aspen, with thick grass on the canyon floors. The layout is a cattleman's dream. A herd can graze here all winter long, protected from winds by the steep canyon walls, feeding on the canyon grasses. Sam and I linger on the point long after I put the cameras away. Then it's time to get back for Rosalee's supper.

When I tell friends from the city about the wonderful people I've met on ranches

Branding calves out on the range.

Wild turkeys are just some of the abundant wildlife that co-exist on the ranch with the cattle.

throughout the West, some respond, "Yes, it must be interesting to see how these simple people live out in the sticks." Nothing could be further from the truth. These are a complex, skilled people. Among Sam and Rosalee's grandchildren, for example, are three who graduated at the top of their classes from Texas A&M, University of Alabama, and Notre Dame. These families are molded daily by the demands of hard work, strategic planning and teamwork — and encouraged by prayer.

After Sam's rattlesnake attack, he hovered near death. But two days later, groggily aware of dark figures hovering above him, Sam heard the pleas calling him back. Catholic nuns were praying over his body, asking God to spare his life. Those prayers and others over the years have woven a protective covering over Sam, Rosalee, the family, and the Pasamonte Ranch.

Upper left: One of the original homesteads that make up part of the Pasamonte Ranch.

The summertime provides good grazing for the cattle on the open plains and in the winter the cattle are protected from the bitter weather in deep canyons. Sam is quick to remind you, however, that all bets are off if it doesn't rain. The three photographs on the right were created when God did send good rain in a recent spring. Sam called me and said, "Jim, you better come back for more pictures. I've never seen the ranch so green. It looks like Ireland."

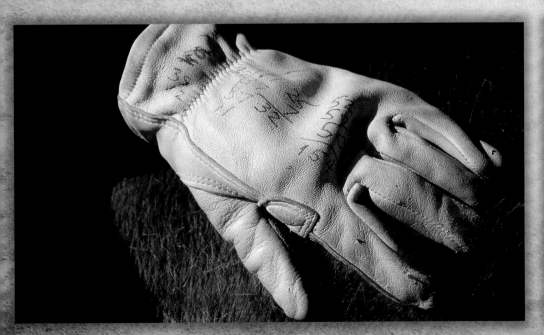

A cowboy's note book.

Scampering up the small hill, my boots slip in sandy soil. The silhouette of a barbed-wire fence looms just ahead of me in the predawn light. Twenty yards on the other side of the fence lay the best point of view for the sunrise. As I work my tripod and cameras through the fence, a barb snags my shirt. Whoever invented this wicked wire has received more than his share of curses, I am sure. Squirming flat on my belly under the lowest wire, I imagine the passing birds laughing at this sight. Dusting off, I set up my equipment. Soon the clouds and fields come to life with glorious, glowing light. I'm grinning with excitement as I change cameras and lenses to record the beauty before me. I remove my glasses and put them in my shirt pocket as I rush to capture the intensifying colors.

Sometimes the hues of a sunrise or sunset last only moments. Other times the colors build and transform for half an hour or more. This is one of those longer-lasting sunrises, and if I hurry there will be time to get back to the car and create photographs at two more locations before meeting the cowhands in the shop at six-thirty. Arriving at the third location, I reach for my glasses. They're not in my pocket. There is no time to search through the camera cases because my perfect light is almost gone. But I can't see the tiny exposure meter on my equipment without the glasses. So I squint and guess at the exposure setting and bracket it with a few settings over and under my estimate.

The eyeglasses must have fallen out of my pocket as I crawled under the barbed wire. I hurry back and try to pinpoint my two earlier locations, but finding these glasses is going to be hopeless. The high grass hides everything below it, and I'm unsure of exactly where I was.

The futile search leaves me worried during the cowhands' work-planning session. I'm preoccupied at possibly being forced to guess all day at the proper exposure settings. And, I admit, I'm embarrassed since the "city slicker" syndrome has struck me once again. But two of the cowboys, Dan and Bryan, take pity and offer to help look for the glasses. They ride their ATVs and I drive my vehicle in the vague direction of where I think I first took pictures this morning. Most all the hands ride ATVs here on the Circle Dot Ranch, a modern twist on "saddling up." True to their roots, though, they do ride their horses when rounding up livestock in the area's canyons and creek bottoms.

Like Native American scouts of old, Dan and Bryan manage to track my footprints precisely to the spot where I set my tripod. No glasses. Then I make a guess at the second location. All the grass and barbed-wire fencing look exactly the same to me, but these guys find the spot, and we scour the ground.

I'm ready to quit looking. "We're never going to find glasses out here without a miracle," I announce. As I head toward my car to search the last location, there's nothing to lose, so I pray for a miracle. "God, probably nobody's ever stood here and prayed before, but please direct us to my glasses." Somehow praying in the company of cowboys seems either very spiritual or stupid.

Suddenly, there's a whoop and a holler: "Jim! Hey, Jim!" the cowboys shout as they both spot my glasses in the deep grass at the same time. Wow! God really does look after old men and blind photographers.

James Monahan was named the Nebraska cattleman of the year in 2005.

MONAHAN RANCH

NEBRASKA

The Sandhills of north central Nebraska are one of the last areas settled in the United States of America. The soil isn't conducive to growing any crops, so farmers never homesteaded the region. During the second half of the 19th century, the hills were rife with renegade Indian bands and outlaws. One of the notorious local outlaws was Doc Middleton, otherwise known as King of the Horse Thieves. Doc's "Pony Boys" supposedly only stole horses from the government in order to assist settlers down on their luck—a sort of Sandhills Robin Hood. And there were many families down on their luck in this stark, unpopulated region. Pioneers bypassed this part of Nebraska in their westward migration until virtually every other territory west of the Mississippi had been claimed.

In 1887 fourteen-year-old James H. Monaghan traveled here from Iowa with his grandfather. James' widowed mother and two sisters later followed. They were a practical people who chose, for example, to change the spelling of their name simply to "Monahan." Three generations later, the current James Monahan says that his grandfather crossed the Missouri and "dropped the G in the river."

While still a teenager, James H. bought a quarter-section of land in 1893 and raised a few cattle, selling beef to the railroad crews for five cents a pound. His boldest venture came when, as a 23-year-old, James rode to Wyoming, bought 600 head of cattle, and drove them himself back to his little ranch. The drive took an entire month. That determination soon saw James purchasing 1,000 acres of land, and in 1923 the Circle Dot brand was born. His grandson James now comments, "That brand was the best thing that ever happened for the ranch. You don't have to worry about a tired cowhand putting it upside down!"

The location of the Circle Dot ranch house itself demonstrates the cleverness of the elder James. The ranch stretches over two counties, Grant and Cherry. So James surveyed the exact boundary between the counties and built the house just on the Grant County side, since the county seat was only twelve miles away. Had he built on the Cherry County side, the county seat would

have been 100 miles by horseback or over 200 miles by rail!

James H. Monahan's son, Earl, was born on the ranch in 1899. Again, ingenuity and determination reigned in that generation of Monahans. The winter of 1913 wreaked havoc on all the nearby ranches when a series of malicious blizzards buried the plains in so much snow that thousands of livestock froze to death. Earl and a high school friend hired out to the neighbors to uncover the frozen, dead cattle and collect the hides. They engineered a unique skinning process. The boys looped a rope from a horse to the head of the cow carcass, and then used a second horse to attach another rope to the cow's hide. The frozen hide peeled off as the horses pulled in opposite directions. Over several weeks of the big freeze, Earl and his partner brought in hides by the wagonload, enough so that the railroad actually left a car positioned near town so the boys could stack their hides in it.

Today, James and Hil (actually Hildreth, but the Monahans like to keep things simple), together with their two sons and a daughter, run this expansive spread. The ranch employs twelve cowhands, some of whom have been with the

Hil dispenses great food and good stories from her kitchen. She tells me of a three-day trip they took recently on their ATVs. It was much more difficult then she was led to believe it would be. After the second day, someone over heard her say to James, "If you ever do this to me again, I will kill you."

The other lady said , "Boy, you sound serious about that. How do you plan to kill him?"

Hil responded, "It would have to be a slow painful death to make up for this trip. I think first we would spend a full day going from one art gallery to another, and follow that up with a reception in a dusty old museum. That would kill him for sure."

All saddled up and not a horse among them and only one cowboy hat. What is this world coming to?

Monahans for almost 40 years. Actually, one cowboy is nearing retirement and has built a house on the property to stay near the Circle Dot operations.

Spring finds the ranch crew handling calving, inoculations, castration and such. Summer means haying. The prairie only grows grass since the Sandhills are just that: sand. (The Monahans tell of an oil exploration team that dug 3600 feet down to solid rock, and most of that was pure sand, with water and some gravel.) But the grass this prairie grows is rich in nutrients, so the Circle Dot gears up every summer to cut, rake, bale, and put up 20,000 huge round bales of hay—feed for the winter. So what happens during the fall?

Football! What else? Rabid Cornhusker fans, the Monahans barter beef for a prime tailgating spot at the stadium and motor to all the University of Nebraska games in Lincoln.

When your driveway is as long as the Monahan's, you have to come up with yard art that is cost effective.

Now here is real art – depending of course, on your football preferences.

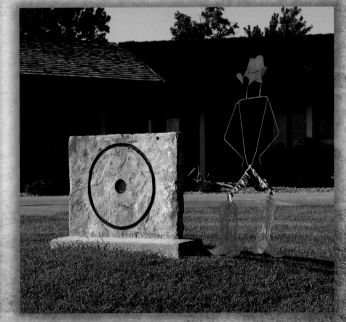

Actually there are three people on horseback that ride the more rugged terrain but the four wheelers can round up the cattle much quicker on the rolling hills.

Déjà vu.

It's 1969, and the snowflakes are winning the war with the windshield wipers as we turn the old Volkswagen onto the dirt road into the cold ghost town of Bodie, California. It's the first storm of the season. The fall colors are still on many of the aspen trees, creating perfect conditions for photography. My college buddy Ed and I talk excitedly about the images we will capture in this ghost town, imagining how the fresh snow will look on the old weathered wood in the morning.

Through the snowfall we spot the ranger's home in one of the restored old buildings. We're the only ghost town visitors on this late fall afternoon, and the ranger greets us from his porch as he pulls on his down parka. I ask him where we can set up camp. Somehow he senses that we're new to winter camping. It must be the faded blue jeans and cotton sweatshirts we are wearing, since cotton is one of the worst fabrics to wear when it's cold or when chances are good that it will get wet. Leading the two of us around to the side of his house, he points out on the large circular thermometer that it is now just twelve degrees.

"Boys, two things can happen here," he says. "This storm can just pass through in a few more hours, and normally that means the temperature will drop way below zero tonight. Or it could just keep on snowing for two weeks solid. If you're prepared for those two possibilities, you can camp just beyond where your car is parked.

Mark Lacey checks on the cattle with the eastern side of the Sierra Nevada mountains in the background. His family has been ranching here since 1870.

LACEY LIVESTOCK

CALIFORNIA

The Panamint Mountains east of the Sierra Nevada Range.

However, if you do that and get snowed in here, don't come to me for help or supplies. If you aren't prepared for these two possibilities, I recommend that you head back down to the highway and check into a cabin."

We're college students. So we're smart enough to know we're dumb about camping in the snow. And we're poor enough to know we can't afford to rent a cabin. The compromise is to camp back down near the highway. Somehow, by wearing all the clothes we brought, we survived a very cold night in lightweight summer sleeping bags and a drafty canvas tent.

It all comes back as I drive up into the high country of the eastern Sierras. Much has changed since my first trip to this area but the land is the same: stark, rough and beautiful. Snowcapped Mount Whitney rises to more than 14,000 feet just to the west, with other "fourteeners" peaking along the Sierra Nevada ridgeline. To the east lie the dry Inyo Mountains. Beyond them, stretching across into Nevada, is the infamous Death Valley desert. Its lowest point is 282 feet below sea level, and its towns are distinctive: Skidoo. Badwater. Then there's Hell, California.

Between the Sierras and the Inyo Mountains, the Owens Valley runs almost north and south—with its more civilized towns of Independence, Lone Pine, and Olancha. At about 4,000 feet in elevation, Olancha is pretty much the center of the Lacey Livestock Company operations—even though Mark and Brenda Lacey live in nearby Independence; Mark's, parents John and Dee, live over the Sierras in Paso Robles; and Mark's sister, Nicki, lives in Hawaii. They lease 50,000 acres of pastureland, running about 1,800 cows and 1,000 yearlings—with another 6,000 grazing on the 200,000 acres of the Tejon Ranch in the San Joaquin Valley. Furthermore, the Laceys work with rancher David Wood in a corporate partnership called Centennial Livestock. Together, the ranchers run yearlings during the winter directly west along the coast in Monterey County. Every year they truck about 4,000 head up to the fertile, open grazing land around Bridgeport in Mono County at the northern end of the Owens Valley.

This is a massive cattle enterprise, so it demands plenty of exceptional horses. In the 1960s, John began breeding the Lacey's own cutting and ranch quarter horses. These award-winning horses are bred specifically for their athletic ability, good looks, and pleasant disposition. John is a former president of the National Cattlemen's Beef Association, and says he also breeds the mares of the Remuda for—there's just no other way to word it—good "cow sense."

In 1867, John's great-grandfather, John William Lacey, traveled from Missouri to prospect for gold and built a homestead in what was then the lush valley near Lone Pine. Neighboring Fort Independence provided protection for the valley's miners and farmers from marauding Indians. In 1915 John W. sold his property and divided the proceeds among his four children, but not before Los Angeles County, nearly 200 miles to the south, had bought up water rights in the entire region.

Son Mark E. Lacey, an ordained Methodist minister, returned to the area in 1923 and purchased land to form what he called the Junction Ranch. Mark fought to grow crops on a 640-acre section using an irrigation network. But even by the mid-1920s, so much of the region's water was diverted to Los Angeles that Owens Lake dried up, and the valley was nearly impossible to farm for lack of water.

Molly and Katie Lacey wait while their Dad gets their horses ready.

Diversifying his income during the austere economy of the 1930s, Mark E. established a gas station, a motel, a store, and a café in Olancha. The off-duty cowboys from his ranch could earn a few extra dollars by doing odd jobs around Mark's businesses in town. Mark built Junction Ranch up to almost 300,000 acres, much of it stretching east into the desert. During World War II, a weapons-testing U.S. Navy base was established to the south of the Lacey property. The base began expanding and confiscated a good amount of the Junction Ranch to form what is now the Naval Air Weapons Station China Lake—the largest land holdings of the U.S. Navy, comprising 1.1 million acres. Mark E. Lacey went to court to contest the seizure of his property, and, amazingly, he won. He received a sizeable settlement from the Department of the Navy that allowed him

Brenda Lacey is the 4H program representative for Mono and Inyo counties.

The Lacey family goes for a late afternoon ride.

to purchase more acreage across the Owens Valley and beyond.

Mark E. Lacey's son John took over the ranch in 1954 and began buying land and running cattle in Oregon, Wyoming, and all over the West. Today, John and his son, Mark, focus their operations in California.

As I photograph the late summer colors and details of this successful family ranch, I find myself wondering: Will it survive in the economic challenges of the future? Then I remember the conversation I had with Mark's wife Brenda. She'd said, "When ranching is in your blood, it's what you'll do. When Mark graduated from college, he didn't go looking for a city job. He didn't go celebrate at the beach or take a cruise. Within three days, he was back on the ranch. He's just a cowboy, through and through." I think that is my answer.

Molly works her horse before a ride.

The Laceys are partners in the Dressler Ranch which has a nicely restored barn originally built before the gold rush in 1842.

My journey to Utah's Redd Ranch takes place the day before Thanksgiving. The drive northward through Monument Valley and along the edge of Canyonlands National Park takes my breath away. These kind of spectacular views don't encourage a photographer to be in a hurry. In fact, the landforms are so amazing that I have to pull over time after time, and begin to wonder how late it'll be when I finally arrive at the ranch headquarters.

At every stop I try to visualize how the early settlers in this region could traverse this severe landscape. I look down deep gorges and up blind canyons. There is no place for a person on horseback to cross this desolate land, let alone a wagon. It's steep, rocky, and dry, pockmarked here and there with dried-up puddles of alkali mud. Where could they get water? It seems impossible. I soon find out how tough it was for the early Redd pioneers.

John Hardison Redd freed his slaves in 1850 and moved from his Tennessee plantation to the Salt Lake Valley in Utah. His son, Lemuel, married two wives and eventually had twenty children. He left both families in the town of New Harmony and joined a Mormon mission group traveling to establish a community called Bluff. This party of 70 families, 400 horses and oxen, and 1,000 head of cattle incredibly trekked a virtually impossible route 150 miles across Utah to the southeast corner of the state. They performed remarkable feats such as building crude roads down steep cliffs by hand. The 83 wagons were sometimes lowered down the drop-off canyon walls and were floated on rafts across the Colorado River. It's significant that not a single life was lost on this trip.

Charles Redd is fixing a broken irrigation pivot. He is the farmer and truck driver for the ranch.

The little community of La Sal and the Redd Ranch Headquarters are right at the base of the La Sal Mountains.

La Sal Cattle Co.

The Redd Ranch is blessed with lots of water rights to streams that run down from the mountains on the north eastern part of the ranch.

Out on the western range there are caves that are still used by sheep herders and cowboys during roundup. Over the years the caves have served as welcome shelter in many storms.

After such a risky undertaking, it's unfortunate that Lemuel Redd wasn't able to stay. Federal anti-polygamy legislation was passed in 1887, and Lemuel was among those indicted. He eluded authorities for several years, but it soon became clear that he would have to find another home. Leaving his first wife and their grown children in Utah, he fled to Mexico with the second family. Getting across the Colorado River again was his biggest hurdle. The family spent the night on a sandbar in the middle of the river, nearly drowning when the water rose as a storm dumped rain far upstream.

When his father departed for Mexico, Lemuel Jr., part of the first family, stayed behind and started the family's ranching operation in southeast Utah. By the 1890s he ran as many as 20,000 sheep in both Utah and Colorado.

But the ranching operation fell on hard times, and was nearing bankruptcy when Lemuel Jr.'s son, Charles, inherited the property. Charles worked hard to pay off creditors within ten years and expanded the Redd Ranch empire to nearly two million acres by 1965. He was something of a Renaissance man by ranching standards, dipping his toes into everything from state politics to international relations to the Boy Scout organization.

A state legislator, Stockman of the Century, and even knighted in 1957 by Queen Elizabeth for his efforts to foster British-American ties, Charles endowed a chair for Western History at Brigham Young University. "I would like somehow," he said of the endowment, "to get into the hearts and souls of young people the lessons of history, particularly those of Western America. The American pioneer has much to teach us, with his insistence on individual freedom of action, his spirit of adventure and his willingness to accept challenge. He reminds us how precious the heritage of individual freedom is. Perhaps more important to youth today is how acceptance of challenge and risk-taking strengthens and contributes to individual growth. Only through the acceptance of great challenges and the struggle with adversity are men's souls enlarged and extended. Learning of the successful settlement of this country, we may gain courage to face squarely the challenges and problems of present-day frontiers."

Today Charles' son Hardy Redd is hosting me for the family's old-fashioned Thanksgiving dinner. His wife Sunny and other family members have been preparing the food for more than a day. The feast begins with breakfast. I get the first serving of the family's pancakes with homemade white syrup, made with buttermilk, vanilla, and other secret ingredients. Hardy says, "If the Mormon Church knew how good this syrup is, they would outlaw it." I'm sure they would.

Only part of the family is able to make it to the ranch this Thanksgiving, but there are still about twenty people around the table. Hardy gives a short Thanksgiving talk before the meal, expressing the family's gratitude for the freedoms we have in the United States and the financial freedom that the family now enjoys with no debt—a recent, wonderful development.

The day after Thanksgiving, Hardy, several family members, and I take a two-hour drive across the winter pastureland to the Needles Overlook. This steep drop-off

Son-in-law, Luis Treviso, has to put down a blind, old, cancer-ridden cow.

Part of the extended Redd family gathers for the Thanksgiving Day meal.

above the Canyonland National Park sits at the western edge of the ranch. They don't need a fence here because the cattle have only to take one look down, and that's enough to keep them from the edge. The technique works for photographers also, and I find myself backing far away from the precipice.

Hardy has turned the ranch operations over to his two sons Charles Jr. and Lowry. Charles is the farmer and trucker, while Lowry is the cowboy. Charles is experimenting with pinto-bean hay, a high-protein source of food that might possibly produce bigger and healthier cattle. The cattle are healthy even now, feeding on nutritious natural grasses grown across the ranch with the plentiful irrigation water streaming down from the La Sal Mountains.

The Redds' involvement in the community is evident everywhere. The La Sel Livestock Company owns the general store in La Sal, and Charles Jr.'s wife Barbara manages it. It's the only place anywhere near here where you can get a few groceries, buy gas, and hear all the latest gossip. Right next door to the store, a building that houses the post office is also owned by the Redd family.

Facing impossible odds from start to finish, every last acre of the Redd Ranch

is thriving. On my way home, I'm mesmerized once more by the towering formations of sandstone cliffs that jut into the sky. I decide this is the perfect place to have spent Thanksgiving this year.

The diversity of landscape on the Redd Ranch ranges from pine and aspen forest to canyon land cliffs. For generations the Redd family have been wise stewards of this land.

Below: One of the "younger" hands gets ready to ride into the mountains.

Left: The work days start early on the ranch, even on the day after Thanksgiving.

Right: To work off the Thanksgiving meal a few of the kids take their horses out for a ride.

The spring rains yielded to hot summer days, and it was the second hunt of the year ordered by the elders. Several six-foot-tall hunters of the Ni-U-Kon-Ska people quietly padded through the prairie grass that reached above their shoulders. These "Children of the Middle Waters"—called Osage by the early French settlers—were skilled traders, bartering the hunt's game for implements and weapons at the mercantile store in town.

The young store employee who met these hunters was Fred Drummond, known to the northeastern Oklahoma Indians as Ts'o-xe. Fred was fluent in the Osage language, and his boss "Old Man Skinner" often let Fred do the bargaining with the Ni-U-Kon-Ska.

Fred's wife, Addie, was a savvy bargainer herself, selling eggs, chickens, and vegetables to the Osage people while she managed the six children in the Drummond household. From the first day of their marriage, Addie literally socked away her trading profits—into one of Fred's socks. When Old Man Skinner eventually decided to sell the business, Fred fumed that he had no way to purchase the store. That, of course, is when Addie presented her husband with the sock, stuffed with over $1,400—when the average U.S. income was about $400 a year. Fred went on to acquire a bigger store in a neighboring town, eventually becoming president of a local bank and mayor of the town.

Fred's son R.C., known by his descendants as "Big Pa-Pa," detested the mercantile business. The story goes that as Big Pa-Pa was loading a wagon for a store customer, he stopped to notice the afternoon's gentle breeze and the bright blue sky above the rolling hills of Oklahoma. He marched back into the shop, tore off his apron, and told his father, "That's it! I've got to be outside the rest of my life. I'm going to be a rancher."

Big Pa-Pa married his childhood sweetheart Kate in 1911. The couple immediately leased 160 acres and bought 40 cows. They were forced to live in a tent for three weeks while the owners moved out of the house. Once in the home, the couple found that hogs liked to crawl around under the flooring to

Ladd Drummond rounding up cattle.

The Drummond Ranch is home to 4,000 wild horses.

DRUMMOND
RANCH

OKLAHOMA

scratch their backs, and horses would rub against the corners of the house, causing it to shake so much Kate feared the walls would fall down. But they didn't, and the Drummond Ranch near Pawhuska, Oklahoma, was firmly established.

In the early 1900's substantial oil reserves were discovered in the region. The Osage tribe as a collective owned the mineral rights throughout the area, and the Ni-U-Kon-Ska grew relatively wealthy. At the same time, the U.S. government pressed the Osage to allot acreage to individual tribe members rather than continue owning the land as a cooperative. With each Osage profiting from his share of oil revenues, the Allotment Act of 1906 resulted in many Indians losing interest in owning any acreage at all since their wealth was in oil. So Big Pa-Pa would buy the unwanted

Ree Drummond with her youngest son, Todd.

Bryce Drummond is led by his mother for a dawn ride.

Alex Drummond is giving a new born calf a ride on her horse as its mother runs ahead.

land, and the Drummond Ranch expanded.

It's an unusually warm winter day as I clamber out of my car at the ranch house. The first Drummonds I meet riding their bikes out front are Big Pa-Pa's great-great-grandkids Alex and Page, ages nine and seven. Their parents, Ladd and Ree, then welcome me to one of the largest ranch operations in all of Oklahoma—and that's saying something! The Drummonds run about 6,400 mother cows in their combined operations. They also handle nearly 4,000 horses. Herds of wild horses are overpopulating and damaging the environment throughout the West, so the government rounds them up and delivers them to the Drummond Ranch, much to the delight of this wandering photographer!

I'm instantly enchanted by this family—by their appreciation of history, their concern for the land, their creativity. For example, Ladd's wife Ree recently launched a modern Little House on the Prairie blog that gets more than 3,000 hits a day. Is she a pioneer woman like the last century's Addie Drummond? "I am nearly a mirror image of her," she says. "Well, except I grew up in the city and never set foot in the country until I met Ladd. And I also have Direct TV, high-speed Internet, air conditioning, toilets, four-wheel drive, and Advil!"

The Drummond Land & Cattle Company is actually split into two properties, one here in the northern Osage area and the other down on the southern border of Oklahoma and Texas, where the Drummonds graze about 12,000 yearlings on wheat. Revenue from the Drummond operations comes strictly from working cows and horses; the oil money that bolsters many Oklahoma ranches is off-limits to the Drummonds since the Osage tribe owns all mineral rights in the region. Ladd and his brother Tim manage all the farming and cattle-raising.

Kim Kill goes after a stray.

I begin to realize what an educated bunch this is as I talk with Ladd and Tim's father, Chuck. Chuck reminisces how his dad always said, "If you don't have the discipline to finish college, you'll not have the discipline to run a ranch." So the Drummonds work smart as well as hard. Chuck notes that many ranchers are reluctant to turn over management to the next generation, often waiting till they're elderly to release control. He says, "I want to be a guide and counselor to my kids instead of turning it over the day I die." Chuck's son Tim was given the complete responsibility of running the ranch when he graduated from college in 1989. Ladd majored in business finance at Arizona State and returned to the ranch a year later.

I sense that Chuck is zealous about ranching and zealous about life. His energy permeates the culture of the Drummond enterprise. Someone commented,

Wheat fields surround the horse cutting arena on the southern portion of the Drummond Ranch.

"He never knew the shallow end of the pool. He always jumped in the deep end!" Modestly, Chuck says, "You actually need just two things to succeed in ranching: a good wife and a good banker." I'm not sure about the banker, but I learned that Chuck's forty-three-year marriage is one of the touchstones of this family's stability and success.

Chuck started breeding cutting horses in 1988, and the ranch gained a national reputation for producing champions. That led to an interest in cutting horse competition. So the Drummonds are very active in a prestigious network: The Chisholm Trail Cutting Horse Association. The famous Chisholm Trail—the main route for cattle drives from Texas north into Kansas a century ago—crossed the Drummond's southern Oklahoma property. So it was only natural that the group's festive competitions should take place there.

Charles Drummond stands on the Chisholm Trail where it crosses the ranch.

Caleb, Tim, Halle , and Missy Drummond take a break from the cutting horse activities in the arena.

Above: Wild fires are such a threat that the Drummonds formed their own volunteer fire department with several surplus water trucks. Quick response is key to controlling a prairie fire.

Left: A cutting horse demonstration and Josh Sellers, one of the long time cowboys on the ranch, working cattle in everyday life.

When I visit the southern segment of the ranch, a cutting horse show is in progress. I wander like a complete foreigner among this unique sub-culture of cutting horse enthusiasts. Elaborate recreation vehicles and expensive horse trailers circle the Drummond's new seventy-stall barn and a riding arena that seems huge enough for a state fair. Before working on this amazing Great Ranches of the West project, I never would have guessed such extraordinary shows were occurring routinely among the ranch families of America. They work hard — but they play hard, too.

The family tells me that last year they hosted forty shows with more than 5,000 participants, some from as far away as Manitoba, Canada. In 2005 they awarded more than $300,000 in prize money. These horse shows are high visibility and create a lot of interest in the sale of Drummond horses, yet they are only a small part of this vast ranching operation.

I drive off from the ranch and the cutting horse show, enjoying the evening sky of Oklahoma and the smell of dirt. Every vehicle that passes is a pickup whose driver waves. And, as I slow through the nearest town of Waurika, I notice a sign touting the tourist attractions of Jefferson County: the annual county fair, fishing at the lake, the horse shows at the Drummond Ranch, and the annual rattlesnake hunt.

There's nothing like the West.

Chuck Sylvester and Lois Sylvester Blasberg, his daughter are the owners of the Circle Bar Ranch.

I t's cold. I mean really cold, the kind of temperature where my nose hairs freeze as soon as I step outdoors. It's minus twenty degrees this morning on the Circle Bar Ranch. I am thankful that it's a perfectly calm day, one of the few days of the year when the wind isn't blowing here in Wyoming. I have done a fair amount of winter camping, so I know how to dress for this kind of cold, but working the controls on my camera can't be done with heavy mittens. I wear just a thin glove liner on my right hand that will allow me to work the camera properly. When my fingers get too cold, I thrust them back into my pocket for a while.

Ranch manager Cal Hancock tells me it has been colder than this already this winter. He takes it in stride since the cattle need to be fed no matter what the weather conditions. Temperatures can swing from one extreme to another here. Cal remembers one March day several years ago when the temperature went from a minus thirty degrees to sixty above within eight hours.

Today's traveler has to be prudent about venturing out into Wyoming winters. The week before I came on the ranch, businessmen anxious to get home made the decision to fly a small

CIRCLE BAR RANCH

WYOMING

private plane in a snowstorm. They didn't make it, crashing into a mountain on the ranch. Cal and I drive for an hour on ranch roads looking for the wreckage, but the property is so vast we don't find a thing. The closest we come to it are the tracks of the FAA investigators heading cross-country in the snow.

And the snow is deep. Makes you wonder how the early ranchers survived. It's as if they made tougher people in those days—people like Emma Ervay.

Jake and Mattie Ervay finished packing the two wagons in Denton County, Texas. It was 1881, and their destination was the Wyoming Territory. Uncivilized country, for certain. Five-year-old Emma and Josie, age eight, were excited. They kept

telling their homemade dolls what adventures they all would have. The cloth-scrap dolls and an extra pair of warm winter clothes were about all the girls could squeeze into the two covered wagons. It would be a long three-month journey. The girls passed the time singing and teasing each other. Every time they stopped to water the animals, the sisters would hurry to help their father. Any excuse to do something with horses was reason enough.

The first year in Wyoming, Jake worked on the Pick Ranch along the North Platte River. With the birth of a third daughter, Jake soon settled his family onto their own ranch, the JE Ranch, situated on the north side of the Rattlesnake Mountains. For many years the JE Ranch served as the region's post office, which delighted the girls. Their neighbors were always stopping by, and they would be the first to hear any news.

Emma and Josie were sent to St. Mary's High School in Salt Lake City. Then in 1890, fourteen-year-old Emma married Ed Claytor. Four years later, the Claytors filed a homestead claim on the south side of the Rattlesnake Mountains, and this homestead became what is now the Circle Bar Ranch.

Josie also married and had a daughter. Tragically, she, her husband, and daughter all drowned trying to cross the Big Horn River at high-water stage. In 1898, Ed and Emma separated, and Emma—a no-nonsense cowgirl now at age 22—bought the ranch from her ex-husband. She had no children and ran the Circle Bar until she died in 1955.

The ranch passed on to another woman, Emma's niece, and then the niece's daughter, Kay. Kay later married Chuck Sylvester, a man legendary in the West for his twenty-five year tenure as general manager of the Denver Western Stock Show. Chuck is now part owner of the Circle Bar with his daughter Lois, the fourth generation of women to run the ranch. Chuck still winters on his farm a little east and just across the border in Colorado, but spends his summers in Wyoming in this rambling, 110-year-old ranch house.

The house is a multi-bedroom log structure with a sod roof—one of only two remaining sod-roofed log homes in all of Wyoming. From the house you can look down a small valley with a

And I thought it was just my nose hairs that froze in this cold.

In Wyoming, there are times when you feel you could be blown down to Colorado if you didn't hang on.

clear view of all the barns and corrals. In the springtime, wildflowers scramble around the fence posts and race across the pasture. There's a quaint little milk house with an ice-cold stream running through it. It's a tranquil, pastoral setting— deceptively peaceful when it comes to the stories this old house could tell. Here's just one:

During a blizzard in the winter of 1919, Emma left a neighbor's house on horseback. After two hours out in the blinding, wind-blown snow, she returned to the neighbor's home with one side of her face, an arm, and a leg severely frostbitten. Once

the storm eased a bit, Emma headed out again toward her property. After she finally arrived home, the ranch-house stovepipe overheated, igniting a fire in the rafters. In the freezing weather, two hired hands climbed up on the roof and dug down through the sod. Emma balanced on a chair while her cook brought buckets of water, which Emma then handed up through the hole in the ceiling to the men on the roof. The water would splash back on Emma as the men threw it onto the burning rafters. They finally put the fire out, but Emma's piano and livingroom furniture were encased in ice — as was Emma.

I guess there are a couple of lessons in a story like this about the Circle Bar's matriarch. When you live on a remote ranch, if you can't do a job, it won't get done. Outside help of any sort is a long, long ways down the road. The second lesson is clear as I bid goodbye to the ranch manager and drive out into the Wyoming winter: it takes a tough cowgirl to tame this country.

Cal Hancock riding out to check on some cattle.

Near left: A constant winter job is cutting or chopping holes in the ice so that the cattle can get to water.

Right: Over the years many cowboys and cowgirls have been called to dinner by this bell.

Below: One of the Oregon Trail routes passed through this valley just south of the Circle Bar, NT Bar and 7D Ranches. These neighboring ranches are all part of Chuck Sylvester's cattle operation.

Left: Ty Van Norman runs the cattle and horse side of the ranch and his uncle Robin Van Norman (to the right) is the farmer.

Far lower right: Troy Van Norman is a veterinarian in Elko but lives and works on the ranch.

Below: Some of the award winning horses in the Van Norman Remuda (a remuda is a herd of horses).

VAN NORMAN RANCH

NEVADA

The airplane is homebuilt. I find myself wondering if I can fit in more than one of my cameras on this flight. This cockpit gives new meaning to the word snug. As I strap in, pilot-mechanic-farmer Dan Van Norman climbs in the other side, telling me he built this plane in a barn over a seven-year span.

"I worked on it mostly during the cold winters," he says.

Just north of Elko, Nevada, at 6,400 feet, the Van Norman Ranch winters are long and cold, often reaching as low as forty degrees below zero. In some years, the cattle and horses need to receive supplemental feed as early as November because of heavy snowfalls. These winter feedings can last into April when the snow finally melts. So it's reassuring for me to note that Dan has had a lot of time to build the plane.

Today, however, is a beautiful fall day. What better way to get an overview of this property and the whole Independence Valley than to take flight? (My wife will later say, "You flew in a what? Was there duct tape? Baling wire?") Birds screech and dash for cover as Dan taxies down the runway. I feel myself relax. The plane does seem to be well-built, and Dan assures me he's a very safe pilot.

The view below rolls out beneath the plane's wings. Miles of sagebrush are interspersed against the vibrant colors of the valley, a spectacular and vast feast for the eyes…and the camera lens. It's a simultaneously rough and beautiful environment. Winters may be wet and cold, but summers are so dry that Dan's dad, Robin, spends the hot months watching his truck's odometer spin as he hauls water in a 4000-gallon tanker truck to the remote parts of the ranch where there is no irrigation. The Van Norman Ranch is actually made up of five separate properties in this high desert region. All bear imaginative names like "The Meadows" and "Roseberry Ranch." The entire Van Norman property encompasses a somewhat rectangular territory that's about twenty-five miles long by eight miles wide. In the hundreds of square miles of land, the plane plays an important role in every cattle roundup. In the old days it was easy to miss a few cows in the high sage brush and the steep canyons. Now, Dan can spy the strays from the air and radio down their locations to his cousins on horseback.

The cousins are Ty, Troy, and Tilly Van Norman—cowhands through and through. Sometimes they are referred to as "buckaroos," a term that came from the 18th-century Spanish-California vaquero. The distinguished style of that era is still evident in the tack, apparel, horsemanship, and livestock handling on the Van Norman ranch. Troy is a veterinarian in Elko, but lives on the ranch and works alongside his brother and sister as much as possible. Tilly lives on the ranch and rides nearly every day. For years she has been part of an all-girl branding team at the Elko County Fair and Livestock Show. In some roping competitions she is often the only female competing, and as they say in the barren stretches of Nevada, she holds her own just fine.

The ranch was established just over a half-century ago when Charlie and Della Van Norman bought a small homestead called "The Smith Place" near Tuscarora. Charlie had just returned from duty in Europe during World War II, where he served as a decorated captain under the famous General George Patton. Charlie and Della bought two mares and began breeding tough, athletic quarter horses in the 1950s. Sons Bill and Robin followed in the couple's footsteps, taking over the ranch's day-to-day operations

Left: Ty and Ronda Van Norman with their daughter Anna.

Above: Once a week a pastor drives out from Elko and holds a Bible study in Robin and Dianne Van Norman's house.

Above right: Tilly Van Norman relaxes with her guitar. She lives and works on the ranch and can be found on her horse just about every day of the year.

This is the ranch complex where Robin, Dianne and their son Dan live.

when their parents passed on. Today the ranch annually breeds about 50 mares and four or five stallions, horses of such caliber that the American Quarter Horse Association recognized the Van Norman Ranch with a Best Remuda Award in 2001. These valuable and highly-desirable horses have sold individually for more than $17,000. The ranch is known for its horses, yet the business is still beef, running about 1500 head of cattle and an equal number of yearlings. The Van Norman family also farms, cultivating 10,000 acres and baling up to 6,500 tons of hay each year for winter feed. Robin, wife, Dianne, and son, Dan, run the farming operation while Bill and Geri's kids, Ty, Troy, and Tilly, oversee the cattle and horse business.

I like these people, and as I begin to hear their stories, I'm even more impressed. It has been a difficult year. Ty and his, wife, Ronda lost one of their two-year-old twins when the little boy died within a week of contracting pneumonia. Ty, Troy, and Tilly's father, Bill, passed away two months later. Losing a loved one is tough no matter where you live, but if you live far from friends and extended family, the network for support is limited. The family depends on God for strength and comfort in this sometimes harsh country. Ronda finds encouragement when she begins each day before six a.m. with a prayer time held over the phone with a friend in the next valley.

I'm thinking about prayer myself as Dan swings us back toward the runway

in his homemade airplane. I notice for the first time how short the runway is. It's just dirt alongside the gravel road that leads to the ranch complex. Don't most airplane accidents happen upon landing? Why am I, a mountaineer, suddenly afraid of heights? As we zoom in for the touchdown I find myself thinking that, if we crash, hopefully the camera will survive to tell the story of these courageous people.

The little plane shudders and bucks over the dirt runway. But Dan Van Norman, pilot and mechanic extraordinaire, is a man whose performance is as good as his word: he's a very safe pilot. We touch down gently and brake to a stop. For Dan, it's just

another routine flight across the Van Norman Ranch. For me, it's a big relief to be back on solid ground.

The ranch complex painted with Fall color is where Troy Van Norman and his wife, Amanda, live.

The yard art here is from Geri Van Norman's yard. She is Ty, Troy and Tilly's mother.

Gary McCurry, who comes out from town to help every once in a while, points to where they will go to round up some cattle.

Dan Van Norman is the ranch mechanic and a pilot. He built his yellow pride and joy that we flew in to get the aerial photographs of the ranch.

Troy opens a gate to move some horses.

Slices of late afternoon light slant across the dim interior of the century-old pole barn, catching dust in the sunbeams as Andy Groseta and his three adult children Paul, Katy, and Anna stow away their saddles. It has been a long day of herding cattle to winter pasture. Both the horses and their riders will eat well tonight.

In addition to large stacks of alfalfa hay and well-worn riding gear, this barn is full of history. If only the timbers could speak, they would begin their tale with Andy's grandfather, Peter Groseta, Sr. He was a determined Croatian immigrant whose fascination and respect for the high-desert landscape resulted in a series of trades, deals, and hard work that launched one of the most respected ranching operations in Arizona.

With the skills of a logger, Peter drove a team of horses high into the mountains to cut and haul sturdy cypress trees for the barn. Seventy years later, these same timbers support the dreams of Andy, his wife Mary Beth, their children, and their grandchildren at the W Dart Ranch. Even the main ranch house itself, acquired when the family bought a neighboring ranch, has been meticulously restored to its original splendor.

Just after the turn of the century, Peter found work in the copper mines of Jerome. The mine belched continuous black smoke back in those early days, a toxin that slowly poisoned the air, not to mention the surrounding population. When a group of farmers won a lawsuit due to crop damage caused by the sulfur smoke, the mining company bought out nearly all the farms and ranches in the area, put a smoke easement—a deed restriction—on them, and sold them back to the farmers and ranchers. This easement prevented any future property owners from suing the copper mine for the smoke it generated. The W Dart Ranch today has a parcel on it that is free of these smoke easements, one of only two such parcels in the Verde Valley.

With Peter's earnings at the mining company, he purchased land for his family farm and ranch in Middle Verde. In the beginning, the Grosetas only had a few head of beef cattle and dairy cows. They also raised sheep, rabbits, chickens, ducks, and turkeys. The family butchered the animals themselves and sold meat to neighbors and miners in Jerome. If someone wanted five pounds of beef, Peter took a saw and carved off that much—not in cuts like steaks or ribs, but in five-pound hunks of meat. The family also raised corn, carrots, cabbage, sweet potatoes, fruit, and eggs to the point of being completely self-sufficient.

W DART
RANCH

ARIZONA

Up through the 1930s, Arizona was open range. Fences were rare, so ranchers let their cattle run with the wild horses that roamed the country. At that time, wild horses were a big problem, and the U.S. Forest Service would round up any unbranded horses, slaughter them, and then grind them for fish food at local hatcheries. Andy recalls that his father, Pete Jr., adopted and broke a wild mare that was appropriately named "Fishy."

Old Peter Groseta's wise purchases and land trades a century ago, along with those of Andy's parents, Pete and Katherine, are the W Dart's legacy today. Andy frowns, however, about one deal that his grandfather passed up. "It was for the land that's now in the heart of Sedona," he tells me. "To old Peter, it was just plain red dirt 'not fit to run a cow on.'" Now, of course, Sedona is well-known as one of the West's most expensive resort towns.

The ranch holdings that the Grosetas have acquired over the years are beautiful, with miles of the Verde River and meandering streams, rocky foothills, and elevations up to 6,000 feet. And yet the W Dart is just an hour-and-a-half drive from Phoenix. This proximity to the burgeoning population of northern Arizona has brought unsettling problems to the Grosetas. Determined vandals steal heavy ranch equipment such as lock-and-chain-secured generators. Rambunctious ATV enthusiasts leave deep tracks that quickly lead to serious erosion. Some of the four-wheelers zoom straight up hills and spin circles on very fragile land, causing damage that will take a hundred years or more to recover. It's terrible that some people have no respect for the land, be it public or private.

In spite of these challenges, the ranch thrives. With Andy's grown children all trained in agricultural sciences, the family remains strategic and forward-thinking in their range management

The Groseta family stands at the entrance to the barn that Peter Groseta Sr. built severa

generations ago. Left to right: Andy, Mary Beth, Anna, Katy, Gretchen, baby Pete, Paul and Grace.

Katy and Anna work on the ranch during Christmas break from their studies.

Katy Groseta graduated, with her Masters Degree in Agricultural Education, from the University of Florida in May 2007.

practices. A herd will be left in a pasture area until just 30-40% of the grass is grazed, and then the cattle are moved to other pastures with fresh feed. The severe drought of 1996 and the following years challenged the family's expertise as they were forced to sell off cattle. In 2002-2003 some of their herds had to be pastured as far away as Oklahoma until the rains came again.

Implementing a good land stewardship ethic by using approved range management practices is a high priority with the Groseta family. Andy, as his grandfather and father did, wants to leave the rangelands on the ranch in better condition than he found them for the next generation of Grosetas. Andy says, "If you take care of the land, it will take care of you."

The family continues to honor the old traditions. I was amused that an historic "sheep driveway" that crosses the ranch is still on the books, a throwback to the days of deep-rooted friction between Arizona sheepherders and cattlemen. Sheep ranchers were given the right to herd their flocks across rangeland and up into the northern Arizona mountains. But to keep that stock

The last light of day shines on the San Francisco Peaks, as the Verde Valley lies in shadow in the foreground. The Grosetas have been ranching in this valley below Flagstaff, Arizona, for four generations.

driveway open, they have to use it at least once annually. So every Spring, sheepherders trail 6,000-8,000 sheep from just north of Phoenix to the high country between Flagstaff and Williams. The trail is a long one, taking three to five days simply to cross the W Dart Ranch.

We prepare to leave the barn, and I take a last look around as the afternoon's winter sunlight turns pale. I wonder how many more Groseta ranchers' dreams this barn will be a part of....

Paul Groseta is a graduate of the University of Arizona and is a ranch real estate appraiser. Of course, he works on The W Dart Ranch too.

Left: Andy, Paul, Katy and Anna Groseta ride through the grass with Don Godard, a long-time employee.

Below that: Andy Groseta, Antonio Landeros and Don Godard

Right: Mary Beth Groseta owns a quilt shop in town. This is the livingroom in the remodeled farm house.

Don Godard used to work for Andy's father and now, though retired, he works for Andy when needed. Here they enjoy a ride across the Verde River.

"Everyone push on the count of three." The mud was nearly knee-deep in places after two days of heavy rain, and the stage wouldn't budge. All the passengers had to climb out, the women standing to the side as the men all pushed. "And three!" The wheels slogged out of the mud with a sucking noise. Women were allowed back on board while the men walked alongside, too muddy to ride inside. Coulson, Montana, began to feel farther and farther away. The driver tried to cheer up the weary travelers: a Pony Express stop lay only a mile ahead, and everyone could spend the night there instead of pushing on to Coulson.

As the driver and brakeman unhitched the horses, the party ate another serving of biscuits and beans, the same old meal served at every stop. Most of the group spent the night sleeping in the barn, since the small Pony Express building was not equipped to host more than a rider or trapper or two. The small talk that night was about Wells Fargo's negotiations to buy this small Montana stage line.

In the morning, the stagecoach was ferried across the Yellowstone River, the longest free-flowing river in the country, and the bedraggled passengers finally reached their destination. Lewis and Clark had traveled through this region in the summer of

VERMILION
RANCH

MONTANA

1806. Clark spent an extra day watering horses here at the mouth of Sweetgrass Creek, impressed by the abundance of wildlife around him.

The site of that old Pony Express stop lies today on the Vermilion Ranch, a grand spread about a dozen miles east of Billings. The ranch was founded in 1867, a full decade before Billings was settled near that little town of Coulson.

The ranch's current owner, Pat Goggins, was born during the Great Depression on a farm in Orland, California—the youngest of six boys. The family's income had plummeted to just $97 annually the year they lost their farm. Pat was four when they packed up and moved to Montana, where the boys'

father worked as a sharecropper. Pat bought the Vermilion in 1961 when it was only a small dairy farm. He now also owns two weekly agricultural newspapers that circulate throughout the West, a feedlot and auction facilities in Billings.

While my wife, Roann, and I wait at the auction yard office for Pat's son-in-law, Bob Cook, I wonder about the huge stuffed head of a bull on one wall. When Bob comes out, I ask him about it. "Oh that's Right Time, a champion bull. We went way out on a limb when we bought him. Cost us $160,000, but Right Time has earned this place of honor. In his lifetime we sold two-and-a-half million dollars' worth of his semen. We shipped it all across the country, into South America, New Zealand and Australia."

Every spring, the Vermilion Ranch breeds about 5,000 cows by artificial insemination. They sell 90% of their bull calves; it's not uncommon for the ranch to sell 600 head of outstanding breeding bulls and over 4,000 commercial Angus heifers at their annual Fall auction. They keep the remaining 10% of bull calves for their own breeding program. Their award-winning Black Angus cattle are raised using no growth hormones at all. To achieve superior specimens, the Goggins rely on good genetics and good feed, most of which comes from their own farming operations on 3,500 irrigated acres where the family grows corn and hay.

I venture out to the back pasture of the Vermilion Ranch to photograph cattle on the site of an old Indian battle called Miller's Fight. The skirmish here occurred six months before the

Battle of the Little Bighorn—Custer's Last Stand—which took place about 60 miles southeast of the ranch. There's a high bluff here, and you can almost sense how the battle scene unfolded.

Reading over my wife's notes later, I find that her words have captured me at work among the cattle. "As Jim walks away from the truck toward the cattle, they scamper away, but not too far. With their glistening black faces looking back, they must wonder what kind of creature this is carrying a strange object on the end of three sticks. Jim sits down behind his tripod and waits. In a few minutes the braver cattle

edge back for a closer look. Soon, curious calves and their mothers surround Jim. The silence is broken every now and then by some bellowing. It echoes off the cliffs like an antiphonal choir. Jim begins making exposures with two cameras—one for close-ups and the other for a wide-angle view. The cattle seem to be posing for him. It's like they know they are being caught on film and are just hoping that they'll become 'cattle stars' and end up being world-famous."

The Vermilion is the Goggins' headquarters, but is only one of several ranches the family owns. The Diamond Ring Ranch runs commercial Angus cattle

on 55,000-plus acres along the Yellowstone River. The Pryor Creek Ranch has 30,000 acres of dry-land ranching. If you watch any Western movies, you've seen the Pryor Creek Ranch—the production location for fourteen major movies—some of them starring Marlon Brando, Liz Taylor, Tom Cruise, and Nicole Kidman, just to drop a few names. Whenever a movie is in production here, the cast and film crew often number in the hundreds, so it's a tremendous boost to the local economy.

Speaking of hundreds of visitors, the Goggins are experts at hospitality. If you think having two or three couples over to your house for dinner entails tons of preparation, consider that the Goggins recently entertained 700 guests for dinner on the ranch. They've hosted that many and more for big barbecue celebrations among their rancher friends. Even during our short stay, Roann and I feel refreshed in the hospitality of the Vermilion.

Though the stagecoaches have long since vanished, a weary traveler can still look to this spot nestled in the verdant Yellowstone River Valley and see more than just a ranch. It's the heart of a family's home and a refuge for those who have journeyed a long way.

James Whittaker separating cattle.

My wife Roann and I drive out of Salt Lake City bound for Leadore, Idaho, population 90. The mountain scenery is stunning as we find ourselves on one of those great no-traffic country roads between Idaho's Sawtooth Range and the Bitterroots of Montana.

James Whittaker's family has been here since 1915, when his grandfather carved out a ranch in this wilderness. As a teenager, James' father, Floyd, ran a twenty-mile trap line for beaver, bobcats, and coyotes high in the mountains above the ranch. He used the money he made from selling pelts to buy more land and continued to buy property before and after the Great Depression. Each purchase enhanced the ability of the next generation to continue to live and ranch in this Lemhi Valley.

The ranch rises to 10,000 feet in elevation and encompasses more than 100,000 acres, including leased land from the U.S. Bureau of Land Management. In the early days, the Whittakers ran sheep on the ranch but little by little switched over to cattle. In 1967 James' father gave him a contract to build twenty-five miles of fence if he would use the money to buy 100 heifer calves. This gave James a good start in his own cattle business.

We pull into the ranch headquarters at 6,300 feet and quickly put on our jackets because of the early morning chill. Many mornings there is music echoing in this high mountain air. It's James' wife, Paula, practicing; she's the church organist in town.

I meet Chase, the youngest of James and Paula's four grown children. He is about to head off to Utah State for his first year of college. I ask him if he's going to major in agricultural science. "Heck, no!" he says. "I'll continue to learn the best ranching skills available right here from my dad. I'm majoring in finance and will bring that knowledge back to the ranch."

Chase's older brother, Jordan, and his wife, Susan, live on one of the ranches down the road that make up the Two Dot spread. Susan is a city girl from Virginia. She only had a two-week introduction to the ranching lifestyle before their wedding date and move to the country. She learned to drive the tractors and hay balers and loves it, harvesting well into the night.

James tells me that his daughter, Jill, studied to be a teacher. When he picked her up from college after graduation, Jill was unusually quiet on the ride

TWO DOT
RANCH

IDAHO

home. James finally asked what was bothering her, and she said, "You know, Dad, I don't want to be a teacher. What I'd really like to do is drive big trucks and heavy equipment." Jill married Boyd Foster, owner of an excavating company and continues to reside nearby, living out her new dream of working the land.

James and Paula's oldest daughter, April, practices as a Montana attorney and owns a small business with her husband, Brian, but still makes time to lend a hand on the ranch whenever needed. April has an unusual birthday tradition. As a present

to herself every year, she spends a week to ten days assisting with calving on the ranch, taking time off from the dog-eat-dog world of business and law to bring new life into the world.

Both April and Jill have their own branded cattle and stake a claim on a portion of the ranch, in addition to helping out when more hands are needed. It's common for family members who live away from the ranch to return to help, especially during events like calving time. Watching over expectant mother cows goes on night and day.

Jill Foster in one of her husband's trucks.

Jill and Boyd Foster with their son, R. J.

Each new calf needs ear tags and shots, while the new mothers are kept in an outside hospital pen and treated to extra nourishing feed. Calves stay in the barn for the day and the mothers are moved inside at night. Idaho's bitterly cold elevations demand that the calves stay inside a day or two after their births, allowing them to slowly acclimate to wintertime temperatures in this high mountain valley.

When James, his dad and grandfather acquired land, they were always careful to include water rights. As James says, "Besides cattle, you need just a few things to be a successful cattleman—water and good grass." In 1902, the Big Timber Creek Water Right and Diversion project was established, and today provides water rights to several ranchers in the region. Recently a lawsuit was filed to block the delivery of this water, hoping it would help stop grazing on public lands. In a victory for ranchers everywhere, this attempt was defeated in court. It's a sad commentary on our litigious society that valuable time and resources must be spent defending our property and rights.

Early one morning, Roann and I drive to the corrals on a southern part of the ranch to photograph the cowboys bringing in a herd of cattle. I see movement far on the horizon, but the cattle are still 40 minutes away. Is something else up on the ridge? Through my binoculars I see a large herd of grazing elk, a beautiful sight! The wildlife is right at home with the domestic cattle. Later, when the cattle finally come into view, I notice there's a cowboy missing. Then I spot a cowboy hat zooming back and forth, just visible above the tall grass. James has decided to ride his dirt bike on this roundup rather than his horse. Roann and I just have to laugh.

One of the properties the Whittakers recently bought was Old Man Barrow's place. The man had a drinking problem that infuriated his wife. After one cattle sale,

Jordan and Susan Whittaker horsing around.

Old Man Barrow spent much of the proceeds on several cases of whiskey. As you can imagine, she went through the roof. That night when he was drunk, she took all the whiskey far out on the back of the ranch and buried it. She never told anyone where. After they died and the Whittakers bought the place, there have been several treasure hunts—all of which, so far, have proven unsuccessful.

Decades ago, James' dad hired a German immigrant who knew everything about ranching, such as the vanishing art of flood irrigation. He worked for the ranch for 59 years and taught the current generation how to work. He had an old German saying: "Make it easy to do it right." James and his children make it all look easy, although I know it's not. On the Two Dot Ranch, they're doing it just right.

Jordan Whittaker

Chase Whittaker

A hay bailer at work.

Bill White uses two different cell phone companies — one works on one part of the ranch, the other on the rest. He is owner of the oldest family-owned ranch in Texas, established in 1819.

The grass is greener on the other side, which is why two Brahman/Hereford-crossed cows, their hooves sinking in mud, are bawling in the shallows of the Intracoastal Waterway. Bill White wades into the current, relying on nothing more than a little encouragement to move the cattle across water nearly two football fields wide to reach greener pastures. I wait on land, swatting at mosquitoes and setting up the camera. There's a marshy smell on the air as a gray cloud of waterfowl rise from the far banks. It's swimming time at the White Ranch in southeast Texas.

Swimming cattle is a centuries-old practice, unique to the state's oldest family ranch. It's held over from the days when ranch founder and patriarch Taylor White drove cattle to New Orleans across four rivers by way of the Opelousas Trail. "The Mississippi had barges," Bill White tells me, prodding one of the cows forward, "but none of the other rivers did. The cattle swam."

In the 1820's, Taylor White began driving his cattle on a path that retraced the Old Spanish Trail all the way to New Orleans, where he could get $12 per steer as compared to $5 per steer back home in Texas, and soon hordes of other ranchers joined him. Until the Union Pacific Railroad finally linked Houston and New Orleans in 1881, the Opelousas Trail echoed with cattle bellows and marked the land with its hieroglyphics of hoofprints reflecting a century of journeys. It seems strange that short-lived northern cattle drives like the Chisolm and the Goodnight inspired so many of America's cowboy legends and folklore, while the Opelousas ran cattle for nine decades, more than all of its famous cousins, yet remains relatively unknown.

Above: The land along the gulf coast is very marshy.
Right: Like kids pushing to be the first in the pool, the cattle sense that their annual Fall swim is just ahead.

WHITE RANCH

TEXAS

The cows stuck in front of me now are a reminder of the hazards of crossing water. So many cattle drowned on the Opelousas Trail that the town of Beaumont, Texas, where the Neches River wanders along the county line, enacted the "Ordinance to Prevent Nuisances by Swimming Cattle" in August of 1840. The provision levied a $6 fine per drowned cow and a $50 bond before crossing. Drowning wasn't the only peril along this drive; there were also the thick humidity, heat, monsoons, robbers, and loneliness. It's a wonder White and his contemporaries ever made it to New Orleans at all. But the journey must have been worth it, because White's longhorn cattle, healthy and sleek from their marsh diets, made him one of the wealthiest men in Texas.

The grass really is greener on the other side. Current ranch owners Bill and Kathleen White still winter their cattle on a rich, three-mile-wide strip of land parallel to the mainland and bordering the Gulf of Mexico. The herd swims across, a group at a time, every fall. "The older cows remember previous winters, how good and green that grass is on the Gulf side," Bill says. "You have to keep an eye out or some of them will swim over there on their own." One cowboy is appointed sea captain and saddles up on a boat instead of a horse, making sure there's enough space between barge traffic on the waterway to squeeze all the migrating cows through.

The White Ranch, at about four feet above sea level, suffers from a host of issues never faced by northern operations—such as alligators and tropical storms. Mosquitoes are such a plague that every cowboy on the ranch includes an arsenal of insect repellent on his saddle from March to October. I'm slapping at them now, swarms of bugs in the humid air. Then, there's violent weather. During Hurricane Rita, the Whites had to move all their cattle into safer parts of the state. But six generations of ranchers have outlasted the severe conditions served up by a harsh southeastern Texas landscape, making the "Crossed W" brand one of the oldest continuously used brands in Texas.

The Whites' history is intertwined not only with the legend of the Opelousas Drive, but with much of Texas history. Bill tells me that Taylor White came here from the Carolinas in 1819, bringing a small herd of cattle to these marshy wetlands east of Galveston Bay. White's father had willed him the "Crossed W" brand, and by

It looks like the start of an open water triathlon, doesn't it?

There is so much barge traffic on the intercoastal waterway that the cattle can only cross in groups of about 400 at a time. A barge would never be able to slow down or stop in time to avoid the slow swimmers.

This year there was a stronger current than normal and even though these two cows drifted only 20 or 30 yards from their normal exit point, they got stuck in the mud. Bill jumped in boots and all to encourage them ashore. Bill said, "I never had to deal with 'stuck-in-the-mud' cows before."

1840 the cattle census listed ten thousand head bearing this mark. Almost five million unbranded cattle were roaming Texas when Taylor White arrived, Spanish longhorns abandoned when the Catholic missions were abandoned in the late 1700s. A finders-keepers rule in effect meant an enterprising cattleman could quickly add numbers to his herd.

With an exponentially growing number of cattle to monitor, the Whites made history and caused a huge uproar when they installed southeastern Texas' first barbed-wire fence in 1883. Twenty miles of barbed wire at a hundred dollars a mile meant only one thing to the neighbors: other ranchers benefiting from free range on the White's property were now barred. But the barbed wire came out of a respect for the land and for boundaries.

The White family has donated thousands of acres to create the Anahuac and McFaddin Wildlife Refuges and continue to maintain excellent relations with the managers in place at each sanctuary. Crossed W cattle run on about half of the

Above: Robert Guillory checks the herd while Mary White relaxes.

refuge land, helping make it more suitable for wildlife. The cows' hoof action around the drinking ponds stimulates fallen seeds, encouraging plant growth, which in turn attracts more birds. The cattle also help control some of the more unwanted weeds and keep down the growth of non-native plants that were introduced years ago or have been blown in by hurricanes.

The cattle with the Crossed W brand have a Brahman influence. This makes them well suited to the climate on the Gulf coast, and they can resist the mosquitoes better.

This is the medical clinic where the cattle get their shots and are sprayed to keep flies and mosquitoes at bay.

The ranch has operated under five flags: Spanish, Mexican, Confederate, Texan, and American. Whether the challenges have come from government rule, climate extremes, or stuck-in-the-mud cattle, the White family has managed to outlast each uncertain event, making them one of the preeminent ranching legacies in Texas.

Bill has finally encouraged the smallest cow forward, and before too long the straggler catches up with the rest of the group. Another successful swim at White

Bill, Mary, Kye and Kathrine White.

December 20, 1896. Running for his life, the cowboy slipped on frozen mud in the corral. Scrambling to his feet, he sprinted into the barn and shimmied up a ladder into the loft. Ernest Hurd's breath turned to puffs of white in the frigid air as he squinted through gaps between hay bales, gun drawn. Cold sweat trickled between his shoulder blades. Just seconds later, his senior partner, William Nottingham, burst through the barn door. Gunfire split the cold silence.

Gunfights settled many disputes in the old days of the West. And this one was settled quickly as Nottingham lay dying in a pool of blood. It was another twist in the strange story of a partnership gone bad in Colorado's beautiful Vail Valley.

Nottingham had traveled from Iowa to the Rocky Mountain valleys with his wife and five small children in 1882. He'd met Ernest Hurd and a third settler near the town of Redcliff, and the three men proposed a ranching partnership. The trio bought 480 acres, brimming with all the excitement and hope that a new business venture in the Wild West brought. But soon tragedy overshadowed the enterprise when the third partner committed suicide. Nottingham and Hurd began quarreling over rising debt and other business issues. Hurd's solution to these arguments with his partner hung in a holster at his side.

Three years after killing Nottingham, Ernest Hurd married William Nottingham's widow, Angeline. Questions and rumors flared all along the valley, and many felt Hurd got his just reward when he died of smallpox shortly after the wedding.

This left the twice-widowed Angeline alone with her three sons and two daughters. They had little beyond the 480 acres of land and a powerful will to survive. Today, her great-grandchildren tell of Angeline's perseverance, especially in the winter

Vern Albertson puts a jacket on to break the chill of the wind.

This is the barn where Vern Albertson met his wife Susan Nottingham.

Susan Nottingham gets ready to "doctor" some cows.
Right: No matter what the weather, you have to check on the mother cows.

NOTTINGHAM-ALBERTSON RANCH

COLORADO

Vern and his brother, Frank, wonder how the pickup can be replaced with a new truck at today's cattle prices.

Young colts rest in the spring sunshine.
Far right: Bill Nottingham in his remote office – a well-used 4x4 pickup.
Below: Young calves and their mothers come running for breakfast.

Neighbors, Bill and Jill Schlegel feeding cattle in the snow.

storms of the high Colorado Rockies. Angeline tended children and animals while learning to ranch.

Through the decades, the Nottinghams herded sheep in the high meadows of what is now Beaver Creek and Bachelor Gulch. Increasingly, Vail's world-famous ski and tourism industry encroached on their ranching operations. Finally, in 1992, William and Angeline Nottingham's grandson, Bill, accepted an offer for what was left of the ranch. With these proceeds, he purchased 20,000 acres fifty miles to the north in Burns Hole.

A century ago, Jack Burns was a trapper, a mountain man who settled in this valley of wolves, six-foot-high sage, and a few cedars. It was so remote and rugged that even the Indians declined to occupy it. The men and women who followed in Jack's footsteps worked long and hard to clear the rocks and sagebrush. They were tough, independent families with names like Gates and Albertson — people who knew how to live off this mountain land. They cured their own meat, made their own soap, and used bear fat for lard—soaking it in vats of raw potatoes when it turned rancid.

In the high valley, these settlers knew that water would be more valuable than land, so together the Burns Hole families spent nine back-breaking summers digging a nine-mile irrigation ditch by hand to bring water to their cattle and pastures. These ranching neighbors have worked together for generations. The harsh environment

demands that folks look out for each other.

So when Bill Nottingham and his family arrived in Burns Hole and built a large timber barn, all the neighbors gathered for a festive Burns Hole Barn Dance. In the best of barn dance traditions, Bill's daughter, Susan, happened to meet her handsome neighbor, Vern Albertson. It wasn't long before they married. Now two historic Colorado families with roots deep in the mountains are united.

My first visit to Vern and Susan's place comes in late winter as calving is underway. Susan drives up on her ATV with two ranch dogs that have hitched a ride in the back. She's been down at the barn, where several cows will soon birth their first calves. Susan puts an average of 10,000 miles a year on the ATV—checking the cattle, irrigation, fencing, and equipment around the ranch. She tells me she feels a little sick every time she returns to the site of her childhood home in the Vail Valley. These days, a Wal-Mart and Home Depot, along with their massive parking lots, cover the foundation of what used to be the Nottingham ranch house. "For me, it's about family and our history. The best thing I do is work with my dad every day. I'm thankful for that privilege, even though so many memories are paved over now."

Her father, Bill, feels the same. Even in the move from Vail, he had no intention of giving up ranching. It is his life, a year-round, 24/7 job and, he says, a "calling" that his family has followed for five generations. Son-in-law Vern evidences a similar link with his settler family history as, during one of my visits, he introduces me to neighbors who still hitch a team of horses to a wagon every morning and feed their animals the old-fashioned

way—pitching hay from the wagon bed. I watch the process through a lens, thankful for a perfect morning with my camera in hand. Light snowflakes swirl and settle on my beard as the majestic Belgian workhorses pull hay to the cattle.

The next fall, I return to the ranch to take part in the high valley community's annual roundup. Earlier in the summer, as they've done for generations, the neighbors from five ranches herded all their cattle to the high country. Together they graze cattle on the rich native grasses all summer long. Then the whole community, plus

friends and the occasional photographer saddle up, ride into the mountains and drive the cattle back before the snow flies.

At a bend in the Colorado River is a little Baptist church where Sunday services start at seven in the morning, so that the ranchers can still put in a full day of work. The church also serves as a community center, hosting my wife and me in mid-November for a Burns Hole potluck dinner and a "world premier" of sorts: I project and narrate a slide show of the fall's roundup photographs. There is a sense of wonder and thankfulness that my camera has managed to chronicle this valley's way of life.

Just a week later, my own family gathers around our Thanksgiving Day meal and thanks God for this caliber of men and women – the ones who raise the best beef in the world.

The Baptist Church on the Colorado River, Burns, Colorado

Margie Gates turns to talk with a volunteer cowgirl at the roundup.

The view looking northwest from the ranch toward the Flat Tops Wilderness area.

Vern Albertson walks through the high grass on the upper range of the ranch above Burns, Colorado.

When the Bells originally bought this ranch, their first foreman, Val Cason, had worked for the notorious Mexican bandito Pancho Villa. As I settle into the ranch house in this southeastern corner of Arizona, I have a few fleeting thoughts about the old Wild West movies I saw as a boy: the shootouts, galloping posses kicking up dust, and banditos threatening villages. I wonder why no one else is staying with me at the hacienda; it seems to remain vacant unless there are family or friends visiting.

I find out why in the ensuing days. The Wild West still lives in this rugged, rocky Bear Valley area. During peak times, nearly 150 undocumented Mexican workers and drug runners steal across the Bell Ranch nightly. Dan Bell, president and managing partner of the ZZ Cattle Corporation, explains how the area transforms into a different world as the sun goes down.

One morning Dan's father, George, decided to go out and check the pumps at the various ranch wells. He parked his pickup at the first pump. Suddenly a helicopter zoomed over the surrounding boulders. Presuming it was the Border Patrol, Dan's father waved, checked the pump and then drove up a small canyon to check two more pumps. The helicopter followed his route, hovering over him as he stepped from his truck. Again he waved, checked the pumps, and drove back to the hacienda. The helicopter pursued and then descended until it was just a few feet above the courtyard. Immediately a line of patrol cars appeared, skidding into the yard as Dan's father fished out his ID and weakly raised his hands.

The Border Patrol officers told him that when he was checking the first pump, two dozen drug runners were hidden in the brush just beyond the well with their backpacks and guns. They likely would have killed him to steal his truck. But why did the helicopter then track him to the other pumps? Actually, they thought he was the drug dealers' pickup man!

ZZ CATTLE CORPORATION

ARIZONA

The next morning, I take a look at the border that marks the edge of not only the Bell Ranch but also the United States of America. It's a barbed-wire fence. The wire is knocked down so often by illegals that Bell Ranch cowhands work a full day each week to return their Mexican neighbors' cattle across this U.S.-Mexico border and then repair the fence.

If it's not armed drug runners lurking in my thoughts, it's the panthers and jaguars that give me an increasingly uneasy feeling as I bunk alone in the Bell Ranch hacienda. A Humboldt State University researcher has positioned eighty-seven motion-triggered cameras near watering holes and trails to film the habits of the two male jaguars that have been sighted at night around the ranch. Dan's cousin, Scott, vice-president of the ranch corporation, is actually aiding the research and checking cameras during my visit.

This valley has always been dangerous. In March of 1886, the fabled Geronimo—leader of the last fighting force of Native Americans—surrendered near

The Bell Ranch headquarters.

The Mexican border is just south of these hills.

here to the U.S. Cavalry. But he soon changed his mind, and for six months he and his Apache raiding party terrorized the citizens in this region around Nogales.

Early on the morning of April 27, 1886, rancher Arthur Peck and his neighbor Charlie were doctoring an injured bull in the shelter of a canyon two miles from his house. Back at the Peck cabin, Arthur's ten-year-old cousin, Trinidad, peeked out the door to see why the dogs were barking. Spying Apaches in the corral, she screamed to her mother, Petra. But Geronimo's men rushed the cabin, killing Petra and the eleven-month-old baby she held in her arms. The raiders took Trinidad captive.

At the canyon mouth, Charlie saw the Indians first. He leaped on his horse, but the Apaches shot him dead. Before Peck could mount his own horse, he was wounded by gunfire. The raiders beat him severely, stripped him naked, and forced him to walk barefoot back to the cabin where Petra and the baby lay murdered in pools of blood. Somehow during the night, Arthur Peck escaped and ran across this rocky ground to the safety of a nearby settlement. No one knows what happened to Trinidad.

That September, Geronimo surrendered again into the personal custody of the U.S. Cavalry's General Nelson Miles. The government transported Geronimo and 450 Apache men, women, and children to Florida for confinement in Forts Marion and Pickens. In 1894 they were moved to Fort Sill in Oklahoma, where Geronimo eventually died.

Thomas Graham Bell bought the ranch next to Arthur Peck's in 1938, when times were still very tough in southwest Arizona. Thomas' son George has the distinction of having been conceived in the barn where Thomas and his new bride June lived in their early years. Joined by Thomas, Jr. a few years later, the Bell family began to build this hacienda ranch house.

Louis Hall, a promising new architect from the University of Arizona, designed the structure with its unique bell tower. The handcrafted metalwork on the doors and on the gates around the house is striking artistry, the work of a railroad blacksmith hired by Dan's grandfather. The blacksmith's fee? One bottle of whiskey per day. The stone wall around the fireplace of this great house has an unusual history: Its stones are from a wall built long ago in a nearby mine. The rocks were numbered and then reassembled in order to wrap perfectly around the fireplace.

Episodes of the 60's television series *Rawhide* starring the young Clint Eastwood were filmed at this picturesque ranch, which also served as the setting for movies from *Pocket Money* with Paul Newman to The *Last Hard Men* with Charlton Heston. But even movie and TV fame would not be enough to save the ranch when a diminutive fish threatened to compromise the Bells' future.

Some years ago a biologist discovered a small fish in one of the streams on Bell property. The Sonora Chub is prevalent in Mexico but isn't found anywhere north of the border except on the Bell Ranch. Therefore the United States included the fish on its list of endangered species. The government ordered the Bells to keep their cattle out of the stream and started building a substantial bridge over the water at great expense. Most of the cost lay in the challenge of constructing the bridge in pieces offsite and then hauling it to the ranch for placement so as not to disturb the fish's habitat.

Dan Bell is widely recognized in Arizona for his expertise in renewable natural resources. He holds a degree in range management from the University of Arizona and cares passionately about preserving the environment of his family's ranch. So he advised the biologist and bridge design teams that if his cattle were not allowed to eat down the willow shoots around the stream, willow trees would flourish and soak up the limited flow of water, resulting in no stream, and no more Sonora Chub.

He was right. Today, several years later, I marvel at the stand of willow trees towering over the bridge on all sides — and there is no stream.

Dan and Roxanne Bell with their children, Katie and Matt.

But there are plenty of rocks. Dan jokes that his cattle are unusually heavy since "all they have to eat is rocks." One morning before dawn, I set my tripod up on a high point of boulders after struggling through what the locals call a "wait-a-minute bush" since the thorns grab you so firmly. The cowboys are on horseback, herding the cattle to higher ground toward me. Hooves scrape against rock a half-mile away, and two coyotes howl in a dark canyon to the west. Chills run up my back as I think what else might be lurking over the southern hill. I'm glad the darkness is dissipating as a warm sun rises in the east. Even in the Wild West of the Bell Ranch, the morning brings a quiet calm.

This is Mexico, just a few yards from the Bell Ranch. The litter is from people waiting for night to come, when they will cross into the US illegally.

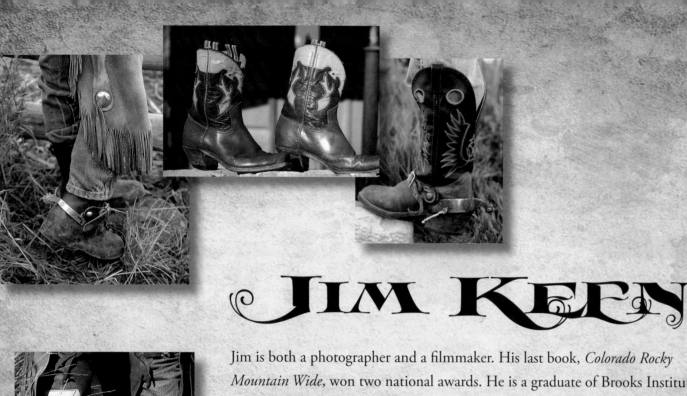

JIM KEEN

Jim is both a photographer and a filmmaker. His last book, *Colorado Rocky Mountain Wide*, won two national awards. He is a graduate of Brooks Institute of Photography and for more than twenty years owned The Air Castle, a photographic gallery in California. Jim has won close to 100 regional and national awards for his photographic art. His work is in private and corporate collections across the US and in fifteen other countries. Professional recognition for his video work includes six Telly awards and three Communicator awards, and his film, *Silent Climb*, was featured at the Breckenridge Film Festival. He lives with his wife, Roann, in Colorado.

For more information and to order enlargements of
your favorite photographs go to:
www.greatrancheswest.com
www.keenmedia.com
1-800/363-5336